Mathematics: The Science of Patterns

Mathematics: The Science of Patterns

The Search for Order in Life, Mind, and the Universe

Keith Devlin

SCIENTIFIC AMERICAN LIBRARY

A division of HPHLP
New York

Library of Congress Cataloging-in-Publication Data

Devlin, Keith J.
     Mathematics, the science of patterns: the search for order in
life, mind, and the universe / Keith Devlin.
          p.   cm.
     Includes bibliographical references and index.
     ISBN 0-7167-5047-3
     ISBN 0-7167-6022-3 (pbk)
     1. Mathematics. I. Title.
QA36.D48 1994   94-11366
510' .1—dc20.   CIP

ISSN 1040-3213

Printed in the United States of America

Scientific American Library
A division of HPHLP
New York

Distributed by W. H. Freeman and Company
41 Madison Avenue, New York, NY 10010
Houndmills, Basingstoke RG21 6XS, England

Second printing 1997

This book is number 52 of a series.

Contents

Preface

This book tries to convey the essence of mathematics, both its historical development and its current breadth. It is not a 'how to' book—it does not try to teach mathematics. It is an 'about' book, that sets out to describe mathematics as a rich and living part of human culture. It is intended for the general reader, and does not assume any mathematical knowledge or ability. Some of the diagrams and formulas may be reminiscent of tedious math classes in school or college, but that is the fault of those classes, not mathematics. A page from *Hamlet* might also bring back unpleasant memories of tedious English literature classes, but that does not make Shakespeare's play any less a classic. The diagrams and the formulas you will find in this book are there to explain, not to train, to help you understand, not to confound.

The idea for this book came from Jerry Lyons in January 1992, during the time he was mathematics editor of W. H. Freeman and Company. It was not until the spring of 1993 that I was able to find the time to start work on the project. By then, Jonathan Cobb had taken over direction of the Scientific American Library series, and it was he who guided the project at Freeman. My colleague Fernando Gouvea read the entire manuscript and offered helpful comments. Doris Schattschneider and Kenneth Millett offered comments on parts of the text. At the Scientific American Library, Susan Moran was the ever-vigilant line editor and Travis Amos obtained the excellent photographs. My family suffered the problems caused by my being in the middle of the project when we moved from Maine to California in order for me to take up the challenge of a new academic position. Sincere thanks to them all.

Historically, almost all leading mathematicians were male. Those days are, I hope, gone forever. This book uses both 'he' and 'she' interchangeably as the generic third-person pronoun.

Keith Devlin   Moraga, California   June 1994

Aside from the correction of a small number of minor errors and a few additions to the list of books suggested for further reading, the only change in this paperback edition is in the account of Fermat's last theorem, which was proved shortly after this book was originally published in 1994 in hardcover. I have changed the text to allow for the discovery and to describe the new proof.

Keith Devlin   Moraga, California   October 1996

Wassily Kandinsky, *In the Black Square*, 1923.

What Is Mathematics?

What is mathematics? Ask this question of persons chosen at random, and you are likely to receive the answer "Mathematics is the study of number." With a bit of prodding as to what kind of study they mean, you may be able to induce them to come up with the description "the *science* of numbers." But that is about as far as you will get. And with that you will have obtained a description of mathematics that ceased to be accurate some two and a half thousand years ago!

Given such a huge misconception, there is scarcely any wonder that your randomly chosen persons are unlikely to realize that research in mathematics is a thriving, worldwide activity, or to accept a suggestion that mathematics permeates, often to a considerable extent, most walks of present-day life and society.

In fact, the answer to the question "What is mathematics?" has changed several times during the course of history.

Up to 500 B.C. or thereabout, mathematics was indeed the study of number. This was the period of Egyptian and Babylonian mathematics. In those civilizations, mathematics consisted almost solely of arithmetic. It was largely utilitarian, and very much of a 'cookbook' variety. ("Do such and such to a number and you will get the answer.")

From 500 B.C. to 300 A.D. was the era of Greek mathematics. The mathematicians of ancient Greece were primarily concerned with geometry. Indeed, they regarded numbers in a geometric fashion, as measurements of length, and when they discovered that there were lengths to which their numbers did not correspond (the discovery of irrational lengths), their study of number largely came to a halt. For the Greeks, with their emphasis on geometry, mathematics was the study of number *and shape.*

In fact, it was only with the Greeks that mathematics came into being as an area of study, and ceased being a collection of techniques for measuring, counting, and accounting. Greek interest in mathematics was not just utilitarian; they regarded mathematics as an intellectual pursuit having both aesthetic and religious elements. Thales introduced the idea that the precisely stated assertions of mathematics could be logically proved by a formal argument. This innovation marked the birth of the theorem, now the bedrock of mathematics. For the Greeks, this approach culminated in the publication of Euclid's *Elements,* the most widely circulated book of all time after the Bible.

There was no major change in the overall nature of mathematics until the middle of the seventeenth century, when Newton (in England) and Leibniz (in Germany) independently invented the calculus. In essence, the calculus is the study of motion and change. Previous mathematics had been largely restricted to the static issues of counting, measuring, and describing shape. With the introduction of techniques to handle motion and change, mathematicians were able to study the motion of the planets and of falling bodies on earth, the workings of machinery, the flow of liquids, the expansion of gases, physical forces such as magnetism and electricity, flight, the growth of plants and animals, the spread of epidemics, the fluctuation of profits, and so on. After Newton and Leibniz, mathematics became the study of number, shape, *motion, change, and space.*

Most of the initial work involving calculus was directed toward the study of physics; indeed, many of the great mathematicians of the period are also regarded as physicists. But from about the middle of the eighteenth century there was an increasing interest in the mathematics itself, not just its ap-

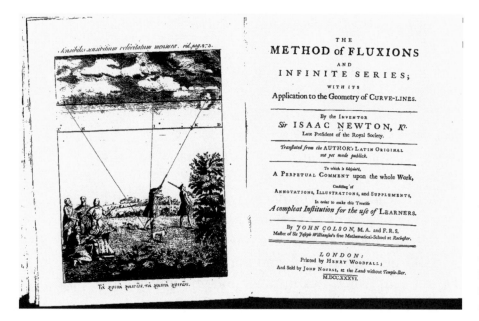

The first calculus textbook. Isaac Newton wrote this account of his method for the analysis of motion using "fluxions" and infinite series in 1671, but it was not published until 1736, nine years after his death.

plications, as mathematicians sought to understand what lay behind the enormous power that the calculus gave to humankind. By the end of the nineteenth century, mathematics had become the study of number, shape, motion, change, and space, *and of the mathematical tools that are used in this study.*

The explosion of mathematical activity that has taken place in the present century has been dramatic. In the year 1900, all the world's mathematical knowledge would have fitted into about eighty books. Today it would take maybe 100,000 volumes to contain all known mathematics. This extraordinary growth has not only been a furtherance of previous mathematics; many quite new branches of mathematics have sprung up. At the turn of the century, mathematics could reasonably be regarded as consisting of some twelve distinct subjects: arithmetic, geometry, calculus, and so on. Today, between sixty and seventy distinct categories would be a reasonable figure. Some subjects, like algebra or topology, have split into various subfields; others, such as complexity theory or dynamical systems theory, are completely new areas of study.

Given this tremendous growth in mathematical activity, for a while it seemed as though the only simple answer to the question "What is mathematics?" was to say, somewhat fatuously, "It is what mathematicians do for a living." A particular study was classified as mathematics not so much because of *what* was studied but because of *how* it was studied—that is, the methodology used. It was only within the last twenty years or so that a definition of mathematics emerged on which most mathematicians now agree: mathematics is *the science of patterns.* What the mathematician does is examine abstract 'patterns'—numerical patterns, patterns of shape, patterns of motion, patterns of behavior, and so on. Those patterns can be either real or imagined, visual or mental, static or dynamic, qualitative or quantitative, purely utilitarian or of little more than recreational interest. They can arise from the world around us, from the depths of space and time, or from the inner workings of the human mind.

To convey the modern conception of mathematics, this book takes six general themes, covering patterns of counting, patterns of reasoning and communicating, patterns of motion and change, patterns of shape, patterns of symmetry and regularity, and patterns of position (topology).

One aspect of modern mathematics that is obvious to even the casual observer is the use of abstract notations: algebraic expressions, complicated-looking formulas, and geometric diagrams. The mathematician's reliance on abstract notation is a reflection of the abstract nature of the patterns she studies.

Different aspects of reality require different forms of description. For example, the most appropriate way to study the lay of the land or to describe to someone how to find their way around a strange town is to draw a map. Text is far less appropriate. Analogously, line drawings in the form of blueprints are the appropriate way to specify the construction of a building. And musical notation is the most appropriate medium to convey music, apart from, perhaps, actually playing the piece.

In the case of various kinds of abstract, 'formal' patterns and abstract structures, the most appropriate means of description and analysis is mathematics, using mathematical notations, concepts, and procedures. For instance, the symbolic notation of algebra is the most appropriate means of describing and analyzing general behavioral properties of addition and multiplication.

For example, the commutative law for addition could be written in English as:

When two numbers are added, their order is not important.

However, it is usually written in the symbolic form

$$m + n = n + m.$$

Such is the complexity and the degree of abstraction of the majority of mathematical patterns, that to use anything other than symbolic notation would be prohibitively cumbersome. And so the development of mathematics has involved a steady increase in the use of abstract notations.

But for all that mathematics books tend to be awash with symbols, mathematical notation no more *is* mathematics than musical notation *is* music. A page of sheet music *represents* a piece of music; the music itself is what you get when the notes on the page are sung or performed on a musical instrument. It is in its performance that the music

DIOPHANTI
ALEXANDRINI
ARITHMETICORVM
LIBRI SEX,
ET DE NVMERIS MVLTANGVLIS
LIBER VNVS.
CVM COMMENTARIIS C. G. BACHETI V. C.
& obseruationibus D. P. de FERMAT Senatoris Tolosani.
Accessit Doctrinæ Analyticæ inuentum nouum, collectum
ex varijs eiusdem D. de FERMAT Epistolis.

TOLOSÆ,
Excudebat BERNARDVS BOSC, è Regione Collegij Societatis Iesu.
M. DC. LXX.

The first systematic use of a recognizably algebraic notation in mathematics seems to have been made by Diophantus, who lived in Alexandria some time around 250 A.D. His treatise *Arithmetic*, of which only six of the original thirteen volumes have been preserved, is generally regarded as the first 'algebra textbook'. In particular, Diophantus used special symbols to denote the unknown in an equation and to denote powers of the unknown, and he employed symbols for subtraction and for equality. The photograph shows the title page of a seventeenth-century Latin translation of Diophantus' classic text.

comes alive and becomes part of our experience; the music exists not on the printed page but in our minds. The same is true for mathematics; the symbols on a page are just a representation of the mathematics. When read by a competent performer (in this case, someone trained in mathematics), the symbols on the printed page come alive—the mathematics lives and breathes in the mind of the reader.

Given the strong similarity between mathematics and music, both of which have their own highly abstract notations and are governed by their own structural rules, it is hardly surprising that many (perhaps most) mathematicians also have some musical talent. And yet, until recently, there was a very obvious difference between mathematics and music. Though only someone well trained in music can read a musical score and hear the music in her head, if that same piece of music is performed by a competent musician, anyone with a sense of hearing can appreciate the result. It requires no musical training to experience and enjoy music when it is performed.

But for most of its history, the only way to appreciate mathematics was to learn how to 'sight-read' the symbols. Though the structures and patterns of mathematics reflect the structure of, and resonate in, the human mind every bit as much as do the structures and patterns of music, human beings have developed no mathematical equivalent to a pair of ears. Mathematics can only be 'seen' with the 'eyes of the mind'. It is as if we had no sense of hearing, so that only someone able to sight-read music would be able to appreciate its patterns and its harmonies.

The development of computer and video technologies has to some extent made mathematics accessible to the untrained. In the hands of a skilled user, the computer can be used to 'perform' mathematics, and the result can be displayed on the screen in a visual form for all to see. Though only a relatively small part of mathematics lends itself to such visual 'performance', it is now possible to convey to the layperson at least something of the beauty and the harmony that the mathematician 'sees' and experiences when she does mathematics.

Like mathematics, music has an abstract notation, used to represent abstract structures.

Without its algebraic symbols, large parts of mathematics simply would not exist. Indeed, the issue is a deep one having to do with human cognitive abilities. The recognition of abstract concepts and the development of an appropriate language are really two sides of the same coin.

The use of a symbol such as a letter, a word, or a picture to denote an abstract entity goes hand in hand with the recognition of that entity *as an entity*. The use of the numeral '7' to denote the number 7 requires that the number 7 be recognized as an entity; the use of the letter '*m*' to denote an arbitrary whole number requires that the *concept* of 'a whole number' be recognized. Having the symbols makes it possible to think about and manipulate the concept.

This linguistic aspect of mathematics is often overlooked, especially in our modern culture, with its emphasis on the procedural, computational aspects of mathematics. Indeed, one often hears the complaint that mathematics would be much easier if it weren't for all that abstract notation, which is rather like saying that Shakespeare would be much easier to understand if it were written in simpler language.

Sadly, the level of abstraction in mathematics, and the consequent need for notations that can cope with that abstraction, means that many, perhaps most, parts of mathematics will remain forever hidden from the nonmathematician; and even the more accessible parts—the parts described in books such as this one—may be at best dimly perceived, with much of their inner beauty locked away from view. Still, that is no excuse for those of us who do seem to have been blessed with an ability to appreciate that inner beauty from trying to communicate to others some sense of what it is we experience—some sense of the simplicity, the precision, the purity, and the elegance that give the patterns of mathematics their very considerable aesthetic value.

Mathematical Symphonies

With the aid of modern computer graphics, the mathematician of today can sometimes arrange a 'performance' of mathematics, in much the same way that a musician can perform a piece of music. In this way, the nonmathematician may catch a brief glimpse of the structures that normally live only in the mathematician's mind. Sometimes, the use of computer graphics can be of significant use to the mathematician as well. The study of so-called complex dynamical systems was begun in the 1920s by the French mathematicians Pierre Fatou and Gaston Julia, but it was not until the late 1970s and early 1980s that the rapidly developing technology of computer graphics enabled Benoit Mandelbrot and other mathematicians to see some of the structures Fatou and Julia had been working with. The strikingly beautiful pictures that emerged from this study have since become something of an art form in their own right. In honor of one of the two pioneers of the subject, certain of these structures are now called Julia sets.

The picture is a computer image of part of a fascinating mathematical object discovered by Mandelbrot, now named after him as the Mandelbrot set. The Mandelbrot set is an example of a rich class of objects known as fractals.

In his 1940 book *A Mathematician's Apology*, the accomplished English mathematician G. H. Hardy wrote:

> The mathematician's patterns, like the painter's or the poet's, must be *beautiful*, the ideas, like the colours or the words, must fit together in a harmonious way. Beauty is the first test; there is no permanent place in the world for ugly mathematics. . . . It may be very hard to *define* mathematical beauty, but that is just as true of beauty of any kind—we may not know quite what we mean by a beautiful poem, but that does not prevent us from recognising one when we read it.

The beauty to which Hardy was referring is, in many cases, a highly abstract, *inner* beauty, a beauty of abstract form and logical structure, a beauty that can be observed, and appreciated, only by those sufficiently well trained in the discipline. It is a beauty "cold and austere," according to Bertrand Russell, the famous English mathematician and philosopher, who wrote, in his 1918 book *Mysticism and Logic*:

> Mathematics, rightly viewed, possesses not only truth, but supreme beauty—a beauty cold and austere, like that of sculpture, without appeal to any part of our weaker nature, without the gorgeous trappings of painting or music, yet sublimely pure, and capable of a stern perfection such as only the greatest art can show.

Mathematics, the science of patterns, is a way of looking at the world, both the physical, biological, and sociological world we inhabit, and the inner world of our minds and thoughts. Mathematics' greatest success has undoubtedly been in the physical domain, where the subject is rightly referred to as both the queen and the servant of the (natural) sciences. Yet, as an entirely human creation, the study of mathematics is ultimately a study of humanity itself. For none of the entities that form the substrate of mathematics exist in the physical world; the numbers, the points, the lines and planes, the surfaces, the

When to See Is to Understand

The mathematician of today can sometimes make use of computer graphics in order to help understand a particular mathematical pattern. The surface shown in this picture was discovered by David Hoffman and William Meeks III in 1983. It is an example of a so-called (non self-intersecting, infinite) minimal surface, the mathematical equivalent of an infinite soap film. Real soap films stretched across a frame always form a surface that occupies the minimal possible area. The mathematician considers abstract analogues that stretch out to infinity. Such surfaces have been studied for over two hundred years, but, until Hoffman and Meeks made their discovery, only three such surfaces were known. Today, as a result of using visualization techniques, mathematicians have discovered many such surfaces.

Much of what is known about minimal surfaces is established by more traditional mathematical techniques, involving lots of algebra and calculus. But, as Hoffman and Meeks showed, the computer graphics can provide the mathematician with the intuition needed to find the right combination of those traditional techniques. A theoretical result by the Brazilian mathematician Celso Costa, in 1983, established the existence of a new infinite minimal surface, but he had no idea what the new surface might look like, or whether it would have the im-

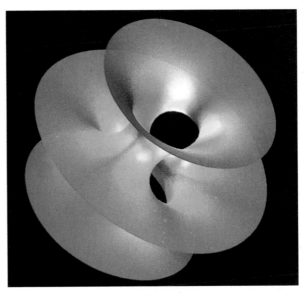

portant property of non self-intersection. Using a new computer graphics package developed by James Hoffman (no relation), David Hoffman and Meeks were able to obtain a picture of the strange new surface. Close examination of the picture enabled them to understand the new surface sufficiently well to develop a proof that it did not intersect itself. They were also able to prove that there were in fact infinitely many non self-intersecting, infinite minimal surfaces.

geometric figures, the functions, and so forth are pure abstractions that exist only in humanity's collective mind. The absolute certainty of a mathematical proof and the indefinitely enduring nature of mathematical truth are reflections of the deep and fundamental status of the mathematician's patterns in both the human mind and the physical world.

In an age when the study of the heavens dominated scientific thought, Galileo said, "The great book of nature can be read only by those who know the language in which it was written. And this language is mathematics." Striking a similar note in a much later era, when the study of the inner work-

ings of the atom had occupied the minds of many scientists for a generation, the Cambridge physicist John Polkinghorne wrote, in 1986, "Mathematics is the abstract key which turns the lock of the physical universe."

In today's age, dominated by information, communication, and computation, mathematics is finding new locks to turn. As the science of abstract patterns, there is scarcely any aspect of our lives that is not affected, to a greater or lesser extent, by mathematics; for abstract patterns are the very essence of thought, of communication, of computation, of society, and of life itself.

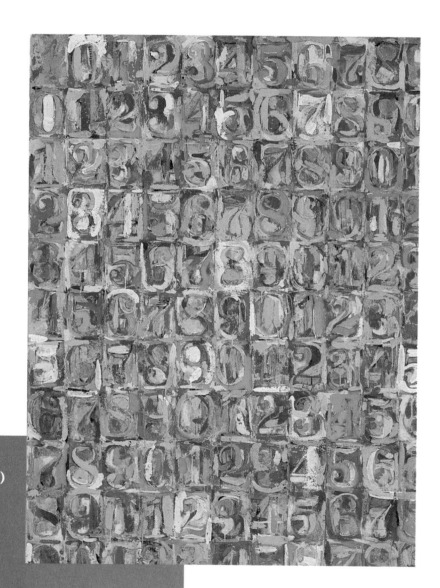

Jasper Johns,
*Numbers in
Color* (1958–59)

1

Counting

Numbers, that is to say, whole numbers, arise from the recognition of patterns in the world around us: the pattern of 'oneness', the pattern of 'twoness', the pattern of 'threeness', and so on. To recognize the pattern that we call 'threeness' is to recognize what it is that a collection of three apples, three children, three footballs, and three rocks have in common. "Can you see a pattern?" a parent might ask a small child, showing her various collections of objects—three apples, three shoes, three gloves, and three toy trucks. The counting numbers 1, 2, 3, . . . are a way of capturing and describing those patterns. The patterns captured by numbers are abstract, and so are the numbers used to describe them.

Having arrived at the number concept as an abstraction of certain patterns in the world around us, another pattern arises at once, a mathematical pattern of the numbers. The numbers are ordered 1, 2, 3, . . . , each succeeding number being greater by 1 than its predecessor.

There are still deeper patterns of number to be examined by the mathematician, patterns of evenness and oddness, of being prime or composite, of being a perfect square, of satisfying various equations, and so forth. The study of number patterns of this form is known as number theory.

The Origins of Number

At the age of about five or less, the typical child in an educated, Western culture makes a cognitive leap that took humankind many thousands of years to achieve: the child acquires the concept of number. He or she comes to realize that there is something common to a collection of, say, five apples, five oranges, five children, five cookies, a rock group of five members, and so on. That common something, 'fiveness', is somehow captured or encapsulated by the number 5, an abstract entity that the child will never see, hear, feel, smell, or taste, but which will have a definite existence for the rest of his or her life. Indeed, such is the role numbers play in everyday life that, for most people, the ordinary counting numbers 1, 2, 3, . . . are more real, more concrete, and certainly more familiar, than Mount Everest or the Taj Mahal.

The conceptual creation of the counting numbers marks the final step in the process of recognizing the *pattern* of 'number of members of a given collection'. This pattern is completely abstract, indeed, so abstract that it is virtually impossible to talk about it except in terms of the abstract numbers themselves. Try explaining what is meant by a collection of twenty-five objects without referring to the *number* 25. (With a very small collection, you can make use of your fingers: a collection of five objects can be explained by holding up the fingers of one hand and saying "This many.")

The acceptance of abstraction does not come easily to the human mind. Given the choice, people prefer the concrete over the abstract. Indeed, work in psychology and anthropology indicates that a facility with abstraction seems to be something we are not born with but acquire, often with great difficulty, as part of our intellectual development.

For instance, according to the work of the cognitive psychologist Jean Piaget, the abstract concept of volume is not innate, but is learnt at an early age. Young children are not able to recognize that a tall, thin glass and a short, stout one can contain the same volume of liquid, even if they see the one poured into the other. For a considerable time, they will maintain that the quantity of liquid changes, that the tall glass contains more than the short one.

The concept of abstract number also appears to be learnt. Small children seem to acquire this concept after they have learned to count. Evidence that the concept of number is not innate comes from the study of cultures that have evolved in isolation from modern society.

For instance, when a member of the Vedda tribe of Sri Lanka wants to count coconuts, he collects a heap of sticks and assigns one to each coconut. Each time he adds a new stick, he says, "That is one." But if asked to say how many coconuts he possesses, he simply points to the pile of sticks and says, "That many." The tribesman thus has a type of counting system, but far from using abstract numbers, he 'counts' in terms of decidedly concrete sticks.

The Vedda tribesman employs a system of counting that dates back to very early times, that of using one collection of objects, say sticks or pebbles, to 'count' the members of another collection, by pairing off the sticks or pebbles with the objects to be 'counted'.

The earliest known man-made artifacts believed to be connected with counting are notched bones, some of which date back to around 35,000 B.C. At least in some cases, the bones seem to have been used as lunar calendars, with each notch representing one sighting of the moon. Similar instances of counting by means of a one-to-one correspondence appear again and again in preliterate societies: pebbles and shells were used in the census in early African kingdoms, and cacao beans and kernels of maize, wheat, and rice were used as counters in the New World.

Of course, any such system suffers from an obvious lack of specificity. A collection of notches, pebbles, or shells indicates a quantity but not the kinds of items being quantified, and hence cannot serve as a means of storing information for long periods. The first known enumeration system that solved this problem was devised in what is now the Middle East, in the region known as the Fertile Crescent, stretching from present-day Syria to Iran.

During the 1970s and early 1980s, anthropologist Denise Schmandt-Besserat of the University of Texas at Austin carried out a detailed study of clay artifacts found in archeological digs at various locations in the Middle East. At every site, among the usual assortment of clay pots, bricks, figurines, and the like, Schmandt-Besserat noticed the presence of collections of small, carefully crafted clay shapes, each measuring between 1 and 3 centimeters across: spheres, disks, cones, tetrahedrons, ovoids, cylinders, triangles, rectangles, and the like. The earliest such objects dated back to around 8,000 B.C., some time after people started to develop agriculture and

first needed to plan harvests and lay down stores of grain for later use.

An organized agriculture required a means of keeping track of a person's stock, and a means to plan and to barter. The clay shapes examined by Schmandt-Besserat appear to have been developed to fulfill this role, with the various shapes being used as tokens to represent the kind of object being counted. For example, there is evidence that a cylinder stood for an animal, cones and spheres stood for two common measures of grain (approximately a peck and a bushel, respectively), and a circular disk stood for a flock. In addition to providing a convenient, physical record of a person's holdings, the clay shapes could be used in planning and bartering, by means of physical manipulation of the tokens.

Clay artifacts like these found in Susa, Iran, were used for accounting in systems of organized agriculture in the Fertile Crescent. *Upper left*: Complex tokens representing *(top row, left to right)* 1 sheep, 1 unit of a particular oil (?), 1 unit of metal, 1 type of garment, and *(bottom row)* 1 garment of a second type, an unknown commodity, and 1 measure of honey. All ca. 3300 B.C. *Bottom left*: An envelope and its content of tokens and the corresponding markings, ca. 3300 B.C. *Above*: An impressed tablet featuring an account of grain, ca. 3100 B.C.

By 6,000 B.C., the use of clay tokens had spread throughout the region. The nature of the clay tokens remained largely unchanged until around 3,000 B.C., when the increasingly more complex societal structure of the Sumerians—characterized by the growth of cities, the rise of the Sumerian temple institution, and the development of organized government—led to the development of more elaborate forms of token. These newer tokens had a greater variety of shapes, including rhomboids, bent coils, and parabolas, and were imprinted with markings. Whereas the plain tokens continued to be used for agricultural accounting, these more complex tokens appear to have been introduced to represent manufactured objects such as garments, metalworks, jars of oil, and loaves of bread.

The stage was set for the next major step toward the development of abstract numbers. During the period 3,300 to 3,250 B.C., as state bureaucracy grew, two means of storing the clay tokens became common. The more elaborate, marked tokens were perforated and strung together on a string attached to an oblong clay frame, and the frame was marked to indicate the identity of the account in question. The plain tokens were stored in clay containers, hollow balls some 5 to 7 centimeters in diameter, and the containers were marked to show the account. Both the strings of tokens and the sealed clay envelopes of tokens thus served as accounts or contracts.

Of course, one obvious drawback of a sealed clay envelope is that the seal has to be broken open in order to examine the contents. So the Sumerian accountants developed the practice of impressing the tokens on the soft exteriors of the envelopes before enclosing them, thereby leaving a visible exterior record of the contents.

But with the contents of the envelope recorded on the exterior, the contents themselves became largely superfluous: all the requisite information was stored on the envelope's outer markings. The tokens themselves could be discarded, which is precisely what happened after a few generations. The result was the birth of the clay tablet, on which impressed marks, and those marks alone, served to record the data previously represented by the tokens. In present-day terminology, we would say that the Sumerian accountants had replaced the physical counting devices by written *numerals*.

From a cognitive viewpoint, it is interesting that the Sumerians did not immediately advance from using physical tokens sealed in a marked envelope to using markings on a single tablet. For some time, the marked clay envelopes redundantly contained the actual tokens depicted by the outer markings. The tokens were regarded as representing the quantity of grain, the number of sheep, or whatever; the envelope's outer markings were regarded as representing not the real-world quantity but the tokens in the envelope. That it took so long to recognize the redundancy of the intermediate tokens suggests that going from physical tokens to an abstract representation was a considerable cognitive development.

Of course, the adoption of a symbolic representation of the amount of grain does not in itself amount to the explicit recognition of the number concept in the sense familiar today, where numbers are regarded as 'things', as 'abstract objects'. Exactly when humankind achieved that feat is hard to say, just as it is not easy to pinpoint the moment when a small child makes a similar cognitive advance. What is certain is that, once the clay tokens had been abandoned, the functioning of Sumerian society relied on the notions of 'oneness', 'twoness', 'threeness', and so on, since that is what the markings on their tablets denoted.

Patterns and Symbols

Having some kind of written numbering system, and using that system to count, as the Sumerians did, is one thing; recognizing a concept of number and investigating the properties of numbers—developing a 'science' of numbers—is quite another. This latter development came much later, when people first began to carry out intellectual investigations of the kind that we would now classify as 'science'.

As an illustration of the distinction between the *use* of a mathematical device and the explicit recognition of the entities involved in that device, take the familiar observation that order is not important when a pair of counting numbers are added or multiplied. (From now on I shall generally refer to counting numbers by the present-day term of *natural numbers*.) Using modern algebraic terminology, this principle can be expressed in a simple, readable fashion by the two commutative laws:

$$m + n = n + m, \qquad m \times n = n \times m.$$

In each of these two identities, the symbols m and n are intended to denote *any* two natural numbers. Using these symbols is quite different from writing down a particular instance of these laws, for example:

$$3 + 8 = 8 + 3, \qquad 3 \times 8 = 8 \times 3.$$

This second case is an observation about the addition and multiplication of two particular numbers.

It requires our having the ability to handle individual abstract numbers, at the very least the abstract numbers 3 and 8, and is typical of the kind of observation that was made by the early Egyptians and Babylonians. But it does not require a well-developed *concept* of abstract number, as do the commutative laws.

By around 2,000 B.C., both the Egyptians and the Babylonians had developed primitive numeral systems and made various geometric observations concerning triangles, pyramids, and the like. Certainly, they 'knew' addition and multiplication were commutative, in the sense that they were familiar with these two patterns of behavior, and undoubtedly made frequent use of commutativity in their daily calculations. But, in their writings, when describing how to perform a particular kind of computation, they did not use algebraic symbols such as m and n. Instead, they always referred to *particular* numbers, although it is clear that in many cases the particular numbers chosen were presented purely as examples, and could be freely replaced by any other numbers.

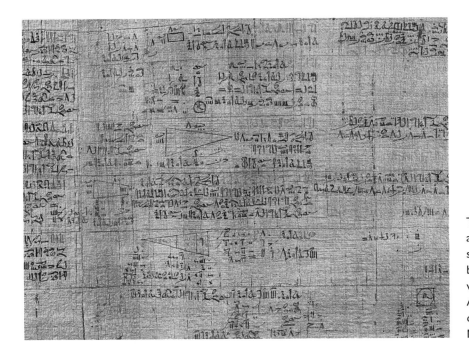

The Rhind papyrus. Written around 650 B.C. by an Egyptian scribe named Ahmes, it is a handbook for performing arithmetic. It was purchased by the Englishman A. Henry Rhind in the nineteenth century, and is now in the British Museum.

For example, in the so-called Moscow Papyrus, an Egyptian document written in 1,850 B.C., appears the following instructions for computing the volume of a certain truncated square pyramid (one with its top 'chopped-off' by a plane parallel to the base):

> If you are told: a truncated pyramid of 6 for the vertical height by 4 on the base by 2 on the top. You are to square this 4, result 16. You are to double 4, result 8. You are to square 2, result 4. You are to add the 16, the 8, and the 4, result 28. You are to take a third of 6, result 2. You are to take 28 twice, result 56. See, it is 56. You will find it right.

Though these instructions are given in terms of particular dimensions, they clearly only make sense as a set of instructions if the reader is free to replace these numbers by any other appropriate values. In modern notation, the result would be expressed by means of an algebraic formula: if the truncated pyramid has a base of sides equal to a, a top of sides equal to b, and a height h, then its volume is given by the formula

$$V = \frac{1}{3}h \, (a^2 + ab + b^2).$$

Being aware of, and utilizing, a certain pattern is not the same as formalizing that pattern and sub-

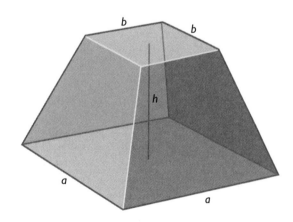

A truncated square pyramid.

jecting it to a scientific analysis. The commutative laws, for example, express certain patterns in the way the natural numbers behave under addition and multiplication, and moreover the laws express these patterns in an explicit fashion. By formulating the laws using algebraic indeterminates such as m and n, entities that denote *arbitrary* natural numbers, we place the focus very definitely on the pattern, not the addition or the multiplication itself.

The general concept of abstract number was not recognized, nor were behavioral rules concerning, say, addition and multiplication formulated, until the era of Greek mathematics began around 600 B.C.

Greek Mathematics

It is not possible to say exactly when 'abstract mathematics' first appeared, but if a time and place had to be set, it would most likely be the sixth century b.c. in Greece, when Thales of Miletus carried out his investigations of geometry. Thales' travels as a merchant undoubtedly exposed him to the known geometrical ideas involved in measurement, but it was apparently not until his own contributions that any attempt was made to regard those geometrical ideas as a subject for systematic investigation in their own right.

Thales took known observations such as

A circle is bisected by any of its diameters

The sides of similar triangles are in proportion

and showed how to deduce them from other, supposedly more 'basic', observations concerning the nature of length and area. The idea of 'mathematical proof' thereby introduced was to become the bedrock of much of mathematics to follow.

One of the most famous early adherents to the concept of mathematical proof was the Greek scholar Pythagoras, who lived some time around 570 to 500 B.C. Few details are known of Pythagoras, since both he and his followers shrouded themselves in mystery, regarding their mathematical studies as something of a black art. He is believed to have been born between 580 and 560 B.C. on the Aegean island of Samos, and to have studied in both Egypt and Babylonia. After several years of wandering, he appears to have settled in Croton, a prosperous Greek settlement in southern Italy. The school he founded there concentrated on the study of *arithmetica* (number theory), *harmonia* (music), *geometria* (geometry), and *astrologia* (astronomy), a fourfold division of knowledge that in the Middle Ages became known as the *quadrivium*. Together with the *trivium* of logic, grammar, and rhetoric, the *quadrivium* made up the seven 'liberal arts' that were regarded as constituting a necessary course of study for an educated person.

Mixed up with the Pythagoreans' philosophical speculations and mystical numerology there was some genuinely rigorous mathematics, including the famous Pythagorean theorem. Illustrated in the figure below, the theorem states that, for any right-angled triangle, the square of the length of the hypotenuse is equal to the sum of the squares of the lengths of the other two sides. This result is remarkable on two counts. First of all, the Pythagore-

ans were able to discern the relationship between the squares of the sides, observing that there was a regular pattern that was exhibited by *all* right-angled triangles. Second, they were able to come up with a rigorous proof that the pattern they had observed did indeed hold for all such triangles.

The abstract patterns of principal interest to the Greek mathematicians were geometric ones, patterns of shape, angle, length, and area. Indeed, apart from the natural numbers, the Greek notion of number was essentially based on geometry, with numbers being thought of as measurements of length and area. All their results concerning angles, lengths, and areas, results which would nowadays be expressed in terms of whole numbers and fractions, were given by the Greeks in the form of comparisons of one angle, length, or area with another. It was this concentration on *ratios* that gave rise to the modern term 'rational number' for a number that can be expressed as a quotient of one whole number by another.

The Greeks discovered various algebraic identities familiar to present-day students of mathematics, such as:

$$(a + b)^2 = a^2 + 2ab + b^2,$$

$$(a - b)^2 = a^2 - 2ab + b^2.$$

Again, these were thought of in geometric terms, as observations about adding and subtracting areas. For example, in Euclid's *Elements* (see presently), the first of these algebraic identities is stated as follows:

Proposition II.4 If a straight line be cut at random, the square on the whole is equal to the squares on the segments and twice the rectangle contained by the segments.

This proposition is illustrated by the left-hand diagram on top of the following page. In this figure, the area of the large square = $(a + b)^2$ = the area of square A + the area of square B + the area of rectangle C + the area of rectangle $D = a^2 + b^2 + ab + ab = a^2 + 2ab + b^2$.

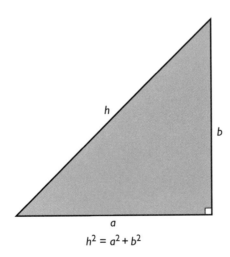

$$h^2 = a^2 + b^2$$

The Pythagorean theorem relates the length of the hypotenuse of a right-angled triangle to the lengths of the other two sides.

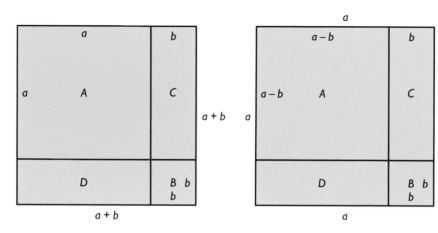

These diagrams show the Greeks' geometric derivation of the algebraic identities for $(a + b)^2$ *(left)* and $(a - b)^2$ *(right)*.

The second identity is derived from the diagram on the right above, in which the sides are labeled differently. In this figure, the area of square $A = (a - b)^2 =$ the area of the large square − the area of the rectangle comprising regions C and B − the area of the rectangles comprising D and B + the area of square B (added on since this area has been subtracted twice, once as part of each rectangle) = $a^2 - ab - ab + b^2 = a^2 - 2ab + b^2$.

Incidentally, the Greek number system did not include negative numbers. Indeed, negative numbers did not come into widespread use until as recently as the eighteenth century.

The Pythagorean theorem can nowadays be expressed by means of the algebraic identity

$$h^2 = a^2 + b^2,$$

where h is the length of the hypotenuse of a given right-angled triangle and a, b are the lengths of the other two sides. The Greeks, however, understood and proved the theorem in purely geometric terms, as a result about the areas of square figures drawn along the three sides of the given triangle. In the figure at the right, area $H =$ area $A +$ area B.

In formulating their results as comparisons of figures, the Greeks were making an assumption that, on occasion, turned out to be far less innocuous than it must have appeared. In modern terminology, they were assuming that any length or area is rational.

The eventual discovery that this particular belief was mistaken came as an immense shock from which Greek mathematics never fully recovered.

The discovery is generally credited to a young mathematician in the Pythagorean school by the

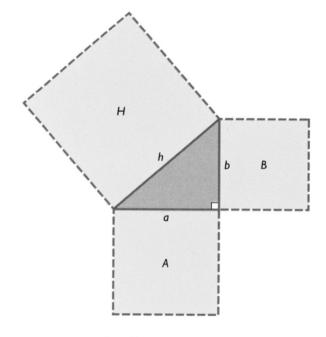

Area $H =$ area $A +$ area B

A geometric proof of the Pythagorean theorem relates the areas of square figures drawn along the three sides of a right-angled triangle.

The Discovery of Irrational Lengths

Theorem There do not exist natural numbers p and q such that $\sqrt{2} = p/q$.

Proof: Suppose, on the contrary, that there were such numbers p, q. If p and q have any common factors, we can cancel them out, so we may as well assume this has already been done, and that p, q have no common factors.

Squaring the identity $\sqrt{2} = p/q$ gives

$$2 = p^2/q^2,$$

which rearranges to give

$$p^2 = 2q^2.$$

This equation tells us that p^2 is an even number. But the square of any even number is even and the square of any odd number is odd. So, as p^2 is even,

it must be the case that p is even. Consequently, p is of the form $p = 2r$ for some natural number r. Substituting $p = 2r$ into the identity $p^2 = 2q^2$ gives

$$4r^2 = 2q^2,$$

which simplifies to

$$2r^2 = q^2.$$

This equation tells us that q^2 is an even number. It follows as in the case of p that q is itself even.

But now we have shown that both p and q are even, which is contrary to the fact assumed at the start that p and q have no common factors. This contradiction implies that the original assumption that such natural numbers p, q exist must have been false. In other words, there are no such p, q. QED.

---

name of Hippasus. He showed that the diagonal of a square cannot be compared to the length of the sides—in modern terminology, the diagonal of a square having rational sides does not have a rational length. Ironically, the proof depends on the Pythagorean theorem.

Suppose that a square has sides 1 unit in length; then, by the Pythagorean theorem, the diagonal has the length $\sqrt{2}$. But, by means of a fairly simple, and extremely elegant, piece of logical reasoning, it can be demonstrated that there are no whole numbers p and q such that $\sqrt{2} = p/q$; the number $\sqrt{2}$ is what mathematicians nowadays refer to as an 'irrational number'. The simple, yet elegant, proof is given in the box on this page.

Such is the power of proof in mathematics that there was no question of ignoring the new result, even though some popular accounts maintain that Hippasus was thrown from a ship and drowned in order to prevent the terrible news from breaking out.

Unfortunately, instead of provoking a search for a system of numbers richer than the rationals—a step that was to come only much later in history, with the development of the 'real numbers'—Hippasus' discovery was regarded as a fundamental impasse.

In between the time of Thales and Pythagoras and the arrival of Euclid on the scene around 330 B.C., Greek mathematics made considerable advances, with the work of Socrates, Plato, Aristotle, and Eudoxus. It was at the Athens Academy, founded by Plato, that Eudoxus worked. There he developed, among other things, a 'theory of proportions' that enabled the Greeks to circumvent, in part, some of the problems created by Hippasus' discovery. And it was also at Plato's Academy that Euclid is reputed to have studied before settling in the new intellectual center of Alexandria, sometime around 330 B.C.

While working in Alexandria at the great Library, the forerunner of today's universities, Euclid

Euclid (ca. 350–300 B.C.), in a fifteenth-century painting by Justus van Ghent.

It is in Books VII to IX that Euclid presents his treatment of what is now known as number theory, the study of the natural numbers. An obvious mathematical pattern exhibited by the natural numbers is that they are ordered one after the other. Number theory examines deeper mathematical patterns found in the natural numbers.

Prime Numbers

Euclid begins Book VII with a list of some twenty-two basic definitions, among which are the following. An *even* number is defined to be one that is divisible into two equal parts and an *odd* number to be one that is not. Somewhat more significant, a *prime* number is defined to be (in modern termi-

produced his mammoth, thirteen-volume work *Elements*. This was a compendium of practically all of Greek mathematics up to the time, containing some 465 propositions from plane and solid geometry and from number theory. Though some of the results were Euclid's own, for the most part his great contribution was the systematic manner in which the mathematics was presented.

Over the centuries since it was written, more than 2,000 editions of *Elements* have been published, and though it contains a number of logical flaws, it remains an excellent example of the 'mathematical method' of commencing with a precise statement of the basic assumptions and thereafter accepting as facts only those results proved from those assumptions.

Books I to VI concentrate on plane geometry and Books XI to XIII deal with solid geometry, both of which are covered in Chapter 4 of this book.

Book X presents an investigation of so-called 'incommensurable magnitudes'. Translated into modern terminology, this volume would be a study of the irrational numbers.

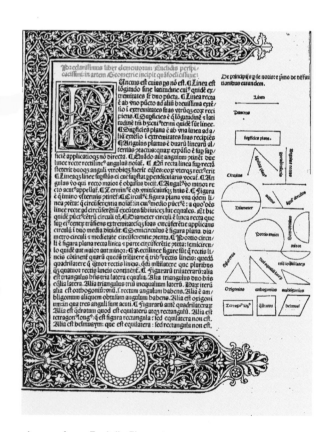

A page from Euclid's *Elements*.

nology) one that has no whole-number divisors other than 1 and the number itself.

For example, among the numbers 1 to 20, the numbers 2, 3, 5, 7, 11, 13, 17, and 19 are primes.

A number greater than 1 that is not prime is said to be composite. Thus, 4, 6, 8, 9, 10, 12, 14, 15, 16, 18, and 20 are the composite numbers in the range 1 to 20.

Among the fundamental results Euclid proved about the primes are the following:

- If a prime number p divides a product mn, then p divides at least one of the two numbers m, n.

- Every natural number is either prime or else can be expressed as a product of primes in a way that is unique apart from the order in which they are written.

- There are infinitely many primes.

The second of these results is of such importance that it is generally referred to as the *fundamental theorem of arithmetic*. Taken together, the first two results tell us that the primes are very much like the physicist's atoms, in that they are the fundamental building blocks out of which all other natural numbers can be built, in this case through the process of multiplication. For example:

$$328,152 = 2 \times 2 \times 2 \times 3 \times 11 \times 11 \times 113.$$

Each of the numbers 2, 3, 11, 113 is prime; they are called the prime factors of 328,152. The product

$$2 \times 2 \times 2 \times 3 \times 11 \times 11 \times 113$$

is called the prime decomposition of 328,152. As with atomic structure, a knowledge of the prime decomposition of a given number can enable the mathematician to say a great deal about the mathematical properties of that number.

The third result, the infinitude of the primes, might come as a surprise to anyone who has spent

The Density of the Primes

N	$\pi(N)$	$\pi(N)/N$
1,000	168	0.168
10,000	1,229	0.123
100,000	9,592	0.095
1,000,000	78,498	0.078
10,000,000	664,579	0.066
100,000,000	5,761,455	0.058

time enumerating prime numbers. Though primes seem to be in great abundance among the first hundred or so natural numbers, they start to thin out as you proceed up through the numbers, and it is not at all clear from the observational evidence whether or not they eventually peter out altogether. For instance, there are eight primes between 2 and 20 but only four between 102 and 120. Going further, of the hundred numbers between 2101 and 2200, only ten are prime, and of the hundred between 10,000,001 and 10,000,100, only two are prime.

The table on this page provides some values of the so-called prime density function, which gives the proportion of primes beneath a given number. To obtain this figure for a given number N, you divide the number of primes less than N, call it $\pi(N)$, by N. In the case of $N = 100$, this ratio works out to be 0.168, which tells you that about 1 in 6 of the numbers below 100 are primes. For $N = 1,000,000$, however, the proportion drops to 0.078, which is about 1 in 13, and for $N = 100,000,000$ it is 0.058, or about 1 in 17. And as N increases, this fall continues.

But for all the steady fall in the ratio $\pi(N)/N$, the primes never peter out completely. Euclid's proof of this fact remains to this day a marvelous example of logical elegance.

The idea is to demonstrate that if you start to list the primes as a sequence $p_1, p_2, p_3, \ldots,$ then this list continues forever. To prove this, you show that if you have listed all the primes up to some prime p_n, then you can always find another prime to add to the list: the list never stops.

Euclid's ingenious idea was to look at the number

$$P = (p_1 \times p_2 \times \cdots \times p_n) + 1,$$

where $p_1, \ldots, p_n$ are all the primes enumerated so far.

If P happens to be prime, then P is a prime bigger than all the primes $p_1, \ldots, p_n$, so the list may be continued. (P might not be the *next* prime after p_n, in which case you will not take P to be $p_{(n+1)}$. But if P is prime, then you know for sure that *there is* a next prime beyond p_n.)

On the other hand, if P is not prime, then P must be evenly divisible by a prime. But none of the primes $p_1, \ldots, p_n$ divides P; if you try to carry out such a division, you will end up with a remainder of 1, that '1' that was added on to give P in the first place. So, if P is not prime, it must be evenly divisible by some prime different from (and hence bigger than) all of $p_1, \ldots, p_n$. In particular, there *must be* a prime bigger than all of $p_1, \ldots, p_n$, so again the sequence can be continued.

It is interesting to observe that, when you look at the number $P_n = (p_1 \times p_2 \times \cdots \times p_n) + 1$ used in Euclid's proof, you don't actually know whether P_n is itself prime or not; the proof uses two arguments, one that works when P_n is prime, one that works when it is not. An obvious question to ask is whether it is always one or the other.

The first few values of P_n look like this:

$$
\begin{aligned}
P_1 &= 2 + 1 & &= 3 \\
P_2 &= (2 \times 3) + 1 & &= 7 \\
P_3 &= (2 \times 3 \times 5) + 1 & &= 31 \\
P_4 &= (2 \times 3 \times 5 \times 7) + 1 & &= 211 \\
P_5 &= (2 \times 3 \times 5 \times 7 \times 11) + 1 &&= 2{,}311.
\end{aligned}
$$

These are all prime numbers. But the next three values are not prime:

$$P_6 = 59 \times 509, \quad P_7 = 19 \times 97 \times 277,$$
$$P_8 = 347 \times 27{,}953.$$

It is not known whether the number P_n is prime for infinitely many values of n. Nor is it known if infinitely many of the numbers P_n are composite. (Of course, at least one of these two alternatives must be true. Most mathematicians would guess that both are in fact true.)

Returning to the density function $\pi(N)/N$, one obvious question is whether there is a *pattern* to the way the density decreases as N gets bigger.

There is certainly no simple pattern. No matter how high up through the numbers you go, you keep finding groups of two or more primes clustered closely together, as well as long stretches that are barren of primes altogether, and these clusters and barren regions seem to occur in a random fashion.

On the positive side, the distribution of primes is not completely chaotic. But nothing was known for certain until well into the nineteenth century, when more sophisticated techniques had been developed. In 1850, the Russian mathematician Pafnuti Chebychef managed to prove that in between any number N and its double $2N$ you can always find at least one prime. So there is *some* order to the way the primes are distributed.

In fact, there is considerable order, but you have to look hard to find it. In 1896, the French mathematicians Jacques Hadamard and Charles de la Vallée Poussin independently proved the remarkable result that, as N gets bigger, the prime density $\pi(N)/N$ gets closer and closer to the quantity $1/\log N$, where $\log N$ is the natural logarithm of N. This result is nowadays referred to as the prime number theorem. It provides a remarkable connection between the natural numbers, which are the fundamental fabric of counting and arithmetic, and the natural logarithm function, which has to do with real numbers and calculus (see Chapter 3). Over a century before it was proved, the prime number theorem had been suspected by the fourteen-year-old mathematical child prodigy Karl Friedrich Gauss. The story recounted in the box on the facing page illustrates the deep mathematical insights Gauss displayed at an even earlier age. The next section describes another of Gauss' major contributions.

Gauss: The Child Genius

Karl Friedrich Gauss (1777–1855).

Born in Brunswick, Germany, in 1777, Karl Friedrich Gauss displayed immense mathematical talent from a very early age. Stories tell of him being able to maintain his father's business accounts at age three. According to another story, while in the elementary school, Gauss confounded his teacher by observing a pattern that enabled him to avoid a decidedly tedious calculation.

Gauss' teacher had asked the class to add together all the numbers from 1 to 100. Presumably the teacher's aim was to keep the students occupied for a time while he was engaged in something else. Unfortunately for him, Gauss quickly spotted the following shortcut to the solution.

You write down the sum twice, once in ascending order, then in descending order, like this:

$$1 + 2 + 3 + \cdots + 98 + 99 + 100$$
$$100 + 99 + 98 + \cdots + 3 + 2 + 1.$$

Now you add the two sums, column by column, to give

$$101 + 101 + 101 + \cdots + 101 + 101 + 101.$$

There are exactly 100 copies of the number 101 in this sum, so its value is

$$100 \times 101 = 10{,}100.$$

Since this product represents twice the answer to the original sum, if you halve it you obtain the answer Gauss' teacher was looking for, namely 5050.

Gauss' trick works for any number n, not just 100. In the general case, when you write the sum from 1 to n in both ascending and descending order and add the two sums column by column, you end up with n copies of the number $n + 1$, which is a total of $n(n+1)$. Halving this total gives the answer:

$$1 + 2 + 3 + \cdots + n = n(n + 1)/2.$$

This formula gives the general pattern of which Gauss' observation was a special case.

It is interesting to note that the formula on the right-hand side of the above identity also captures a geometric pattern. Numbers of the form $n(n + 1)/2$ are called *triangular* numbers, since they are exactly the numbers you can obtain by arranging balls in an equilateral triangle. The first five triangular numbers, 1, 3, 6, 10, 15, are shown below.

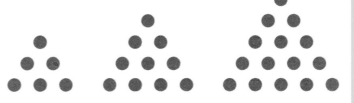

Finite Arithmetic

Finite arithmetic, also known as modular arithmetic, was described in Gauss' book *Disquisitiones Arithmeticae,* an enormously influential work published in 1801, when Gauss was just twenty-four years old.

The idea Gauss investigated is as old as counting. You obtain a finite arithmetic whenever you have a counting system that periodically cycles back on itself and starts again. For instance, when you tell the time, you use a form of finite arithmetic: you count the hours 1, 2, 3, and so on, but when you reach 12 you start over again, 1, 2, 3, and so on. Similarly, you count minutes from 1 to 60 and then start over again.

To turn this familiar concept into a piece of proper mathematics, you have to change the numbering a little, to start counting from 0. With this modification, you would count hours as 0, 1, 2, up to 11, and then the next hour would be 0 again; minutes would be counted off 0, 1, 2, up to 59, and then you would start back at 0.

Though the notion of finite arithmetic was not new, Gauss was the first person to develop it as a significant piece of mathematics, investigating the arithmetic of such number systems. The results are often simple, and occasionally quite startling. For example, in the case of the hours arithmetic, if you add 2 and 3 you get 5 (3 hours after 2 o'clock is 5 o'clock), and if you add 7 and 6 you get 1 (6 hours after 7 o'clock is 1 o'clock). This is familiar enough. But if you write down the sums using standard arithmetical notation, the second addition looks strange:

$$2 + 3 = 5, \qquad 7 + 6 = 1.$$

In the case of the minutes arithmetic, 0 minutes after 45 minutes past the hour is 45 minutes past the hour, and 12 minutes after 48 minutes past the hour is 0 minutes past the hour. These two additions look like this:

$$45 + 0 = 45, \qquad 48 + 12 = 0.$$

For all its strangeness, writing out 'clock arithmetic' in this fashion was a smart move on Gauss' part. It turns out that almost all the rules of ordinary arithmetic are true for finite arithmetic, a classic case of a mathematical pattern carrying over from one area to another (in this case, from ordinary arithmetic to finite arithmetic).

In order to avoid confusing addition and multiplication in finite arithmetic with ordinary arithmetic, Gauss replaced the equality symbol by $\equiv$, and referred to this relation not as 'equality' but as *congruence.* So, the first two arithmetical results above would be written as

$$2 + 3 \equiv 5, \qquad 7 + 6 \equiv 1.$$

The number at which you start over again, 12 or 60 in the two examples considered, is called the *modulus* of the arithmetic. Obviously, there is nothing special about 12 or 60; these are just the values familiar in telling the time. For any natural number n, there will be a corresponding finite arithmetic, the *modular arithmetic of modulus n,* where the numbers are 0, 1, 2, . . . , $n - 1$, and where, when you add or multiply numbers, you discard all whole multiples of n.

I did not give any examples of multiplication above, since we never multiply times of the hour or of the day. But multiplication makes perfect sense from a mathematical point of view. As with addition, you just perform the multiplication in the usual way, but then discard all multiples of the modulus n. So, for example, with modulus 7:

$$2 \times 3 \equiv 6, \qquad 3 \times 5 \equiv 1.$$

Gauss' notion of congruence is often used in mathematics, sometimes with several different moduli at the same time. When this is the case, in order to emphasize the modulus being used on each occasion, mathematicians generally write congruences like this:

$$a \equiv b \; (\text{mod } n),$$

where n is the modulus concerned for this particular congruence. This expression is read as "a is congruent to b modulo n."

For any modulus, the operations of addition, subtraction, and multiplication are entirely straightforward. (I did not describe subtraction above, but it should be obvious how it works. In terms of the two clock arithmetics, subtraction corresponds to counting backward in time.) Division is more of a problem: sometimes you can divide, sometimes you cannot.

For example, in modulus 12 you can divide 7 by 5, and the answer is 11:

$$\frac{7}{5} \equiv 11.$$

To check this, multiply back up by 5, to give

$$7 \equiv 5 \times 11,$$

which is correct, since when you discard multiples of 12 from 55, you are left with 7. But in modulus 12 it is not possible to divide any number by 6, apart from 6 itself. For example, you cannot divide 5 by 6. One way to see this is to observe that if you multiply any of the numbers from 1 to 11 by 6, the result will be an even number, and hence cannot be congruent to 5 modulo 12.

However, in the case where the modulus n is a prime number, division is always possible. So, for a prime modulus, the corresponding modular arithmetic has all the familiar properties of ordinary arithmetic performed with the rational or the real numbers; in mathematician's language, it is a *field*. (Fields appear again on page 54.) And there you have yet another pattern: the pattern that connects the primes with the ability to perform division in modular arithmetic.

Prime Number Patterns

Over the years, a great many mathematicians, both professional and amateur, have looked for patterns in the natural numbers, generally patterns involv-

Pierre de Fermat (1601–1665).

ing primes. Of particular note in this regard is Pierre de Fermat, who lived in France from 1601 to 1665. Fermat was a lawyer by profession, attached to the provincial parliament at Toulouse, and it was not until he was in his thirties that he took up mathematics as a hobby. It turned out to be quite a hobby; in addition to making a number of highly significant discoveries in number theory, he developed a form of analytic geometry some years before René Descartes, founded the subject of probability theory in correspondence with Blaise Pascal, and laid much of the groundwork for the development of the differential calculus, which was to come to fruition a few years later with the work of Gottfried Leibniz and Isaac Newton.

As an amateur mathematician, Fermat published little of his work; what is known of his many accomplishments comes largely from the writings of others, for he made up in letter writing what was not forthcoming by way of publication, maintaining a regular correspondence with some of the finest mathematicians in Europe.

Fermat's skill in number theory was twofold: not only was he able to come up with proofs of some very deep results, he had an uncanny ability to spot number-theoretic patterns in the first place. For instance, in a letter to a correspondent in 1640, Fermat observed that if a is any natural number and p is a prime that does not divide into a, then p has to divide into $a^{p-1} - 1$.

For example, take $a = 8$ and $p = 5$. Since 5 does not divide 8, 5 must divide $8^4 - 1$, according to Fermat's observation. If you work out this number, you obtain $8^4 - 1 = 4,096 - 1 = 4,095$, and indeed you can see at once that 5 does divide this number. Similarly, it must also be the case that 19 divides $145^{18} - 1$, though in this case most people would be justifiably hesitant to try to check the result by direct calculation.

Though perhaps not apparent at first encounter, Fermat's observation turns out to have several important consequences, not only in mathematics but also in other walks of life (among them the design of certain data encryption systems and a number of conjuror's card tricks). In fact, the result crops up so often that mathematicians have given it a name: it is called *Fermat's little theorem*. Nowadays, a number of highly ingenious proofs of this theorem are known, but no one has any idea how Fermat himself proved it. As was his habit, he kept his own methods secret, offering the result as a challenge to others. In the case of his 'little theorem', a complete proof was not found until 1736, when the great Swiss mathematician Leonhard Euler finally rose to meet Fermat's challenge.

Fermat's little theorem can be reformulated in terms of modular arithmetic as follows. If p is a prime, and a is any number between 1 and $p - 1$ inclusive, then

$$a^{p-1} \equiv 1 \pmod{p}.$$

Taking the case $a = 2$, for any prime p greater than 2,

$$2^{p-1} \equiv 1 \pmod{p}.$$

Consequently, given a number p, if $2^{p-1} \not\equiv 1 \pmod{p}$, then p *cannot* be prime.

This formulation of Fermat's result provides an efficient way to try to determine whether a given number is prime. The most obvious method to test if a number N is prime is to look for prime factors. To do this, you may have to trial-divide N by all prime numbers up to $\sqrt{N}$. (You do not need to search beyond $\sqrt{N}$, since if N has a prime factor it must have one no bigger than $\sqrt{N}$.) For fairly small numbers, this is a reasonable approach. Given a moderately powerful computer, the calculations will run virtually instantaneously for any number with 10 digits or less. For instance, if N has 10 digits, $\sqrt{N}$ will have 5 digits, and so will be less than 100,000. So, referring to the table on page 19, there are fewer than 10,000 primes to generate and trial divide into N. This is child's play for a modern computer capable of performing over a million arithmetic operations a second. But even the most powerful computer could take up to two hours to cope with a 20-digit number, and a 50-digit number could require ten billion years. Of course, you could strike it lucky and hit a prime divisor fairly soon; the problem occurs when the number N is prime, since you will then have to test prime factors all the way to $\sqrt{N}$ before you are done.

So, testing if a number is prime by trial division is not feasible for numbers having many more than 20 digits. But, by looking for patterns in the primes, mathematicians have been able to devise a number of alternative means to determine if a given number is prime. Fermat's little theorem provides one such method. To test whether a given number p is prime using Fermat's little theorem, you compute 2^{p-1} in mod-p arithmetic. If the answer is anything other than 1, you know p cannot be prime. But what happens if the answer does turn out to be 1? Unfortunately, you cannot conclude that p is prime. The problem is that, although $2^{p-1} \equiv 1$ (mod p) whenever p is prime, there are also some nonprimes p for which this is true. The smallest such number is 341, which is the product of 11 and 31.

The method would still be fine, provided 341 were just one of a handful of such numbers, since you could check to see if p is one of these awkward ones. Unfortunately, there are infinitely many awkward numbers. Thus, Fermat's little theorem is only reliable for showing that a particular number is *composite*: if $2^{p-1} \not\equiv 1$ (mod p), the number p is definitely composite. When $2^{p-1} \equiv 1$ (mod p), p *may* be prime, but then again it may not. If you are feeling lucky, you might want to take a chance and assume the number is prime anyway, and the odds will be on your side. Composite numbers p for which $2^{p-1} \equiv 1$ (mod p) are fairly rare; there are just two below 1,000, namely 341 and 561, and only 245 below 1,000,000. But, since there are infinitely many such rare numbers altogether, mathematically speaking it is not a safe bet that p is prime, and a far cry from mathematical certainty.

Thus, Fermat's little theorem fails to provide a completely reliable means to test if a number is prime because of the uncertainty that arises when $2^{p-1} \equiv 1$ (mod p). In 1986, the mathematicians L. M. Adleman, R. S. Rumely, H. Cohen, and H. W. Lenstra found a way to eliminate this uncertainty. Starting with Fermat's little theorem, they developed what has turned out to be one of the best general-purpose methods available today for testing whether a number is prime. Known as the ARCL test, it can be run on a fast supercomputer and will take less than ten seconds for a 20-digit number and less than fifteen seconds for a 50-digit number.

The ARCL test is completely reliable. It is referred to as 'general purpose' because it can be used on any number N. A number of primality tests have been devised that work only on numbers of particular forms, such as numbers of the form $b^n + 1$ for some b, n. In such special cases, it may be possible to handle numbers that, in terms of sheer size, would defeat even the ARCL test.

The ability to find large primes became significant outside mathematics when it was discovered that large prime numbers could be used to encrypt messages sent by insecure channels such as telephone lines or radio transmission. The box on the following page provides the details.

Over the years, mathematicians have proposed numerous simple conjectures about prime numbers that, for all their apparent simplicity, remain unresolved to this day.

One example is the *Goldbach conjecture*, raised by Christian Goldbach in a letter to Euler written in 1742. This conjecture proposes that every even number greater than 2 is a sum of two primes. Calculation reveals that this is certainly true for the first few even numbers: $4 = 2 + 2$, $6 = 3 + 3$, $8 = 3 + 5$, $10 = 5 + 5$, $12 = 5 + 7$, and so on. And computer searches have verified the result to at least a billion. But, for all its simplicity, it is still not known for certain whether the conjecture is true or false.

Likewise, no one has been able to settle the question whether every even number can be written as a difference of two consecutive primes in infinitely many ways.

Then there is the unsolved *twin primes conjecture*: are there infinitely many pairs of 'twin primes', primes that are just two whole numbers apart, such as 3 and 5, 11 and 13, 17 and 19, or, going a little higher, 1,000,000,000,061 and 1,000,000,000,063?

Even older than the Goldbach conjecture is a problem posed by Fermat's contemporary Mersenne. In his 1644 book *Cogitata Physica-Mathematica*, Mersenne stated that the numbers $M_n = 2^n - 1$ are prime for $n = 2, 3, 5, 7, 13, 17, 19, 31, 67, 127, 257$, and composite for all other values of n less than 257. No one knows how he arrived at this assertion, but he was not far from the truth: with the arrival of desk calculators, it became possible to check Mersenne's claim, and in 1947 it was discovered that he made just five mistakes; M_{67} and M_{257} are not prime, and M_{61}, M_{89}, M_{107} are prime.

Message Encryption

Using modern computer technology, ingenious primality tests such as the ARCL test make it an easy matter to find prime numbers of, say, 50 to 100 digits. And, if you find two prime numbers, each of, say, 75 digits, your computer can multiply them together to obtain a composite number of 150 digits. Suppose you were now to give this 150-digit number to a stranger, and ask her to find its prime factors. Even if you were to tell the stranger that the number was a product of just two, very large primes, it is extremely unlikely that she would be able to complete the assignment. For while testing a 150-digit number to see if it is prime is a task that a computer can perform in seconds, the best-known means of factoring numbers of that size tend to take impossibly long, years, if not decades or centuries, even on the fastest computer available.

Factoring large numbers is difficult not because mathematicians have failed to invent clever methods for the task. As mentioned in the text, the naive method of trial division can take billions of years for a single 50-digit number, so being able to factor numbers with 80 digits—the current limit— is already an achievement. But whereas the best available primality tests can cope with numbers having a thousand digits, there are no factoring methods known that can begin to approach this performance level, and indeed, there may be no such methods: factoring might be an intrinsically more difficult computational task than testing whether a number is prime.

This huge disparity between the ease of finding large primes and the difficulty of factoring large numbers has been exploited in order to devise one of the most secure forms of 'public key' cipher systems known.

A typical modern cipher system for encrypting messages that have to be sent over insecure, electronic communication channels is illustrated on the facing page. The basic components of the system are two computer programs, an encryptor and a de- cryptor. Because the design of cipher systems is a highly specialized and time-consuming business, it would be impractical, and probably very insecure, to design separate programs for every customer. Thus, the basic encryption/decryption software tends to be available 'off the shelf', for anyone to purchase. Security for the sender and receiver is achieved by requiring a 'key' for both encryption and decryption. Typically, the key will consist of some large number, a hundred or more digits long. The security of the system depends on keeping the key secret. For this reason, the users of such systems generally change the key at frequent intervals.

One obvious problem is the distribution of the key. How does one party send the key to the other? To transmit it over the very electronic communication link that the system is supposed to make secure is obviously out of the question. Indeed, the only really safe way is to send the key physically, by trusted courier. This may be acceptable if only two parties are involved, but completely infeasible for establishing secure communication between, say, all the world's banks and trading companies. In the financial and business world, it is important that any one bank or business be able to contact any other, perhaps at a moment's notice, and be confident that their transaction is secure.

It was in order to meet this kind of requirement that, in 1975, Whitfield Diffie and Martin Hellman proposed the idea of a public key cryptosystem (PKS). In a PKS, each potential message receiver A (which would be anyone who intends to use the system) uses software provided to generate not one but *two* keys, an encryption key and a decryption key. The encryption key is published in a publicly available, on-line directory.

Anyone who wishes to send a message to A looks up A's encryption key, uses that key to encrypt the message, and then sends it off. A then uses her decryption key, which she has divulged to no one, to decrypt the message. Though the basic idea is simple, actually designing such a system is not. The one originally proposed by Diffie and Hellman turned out to be not as secure as they had thought, but a method devised a short time later by Ronald Rivest, Adi Shamir, and Leonard Adleman has proved to be much more robust, and the RSA system, as it is known, is now widely used in the international banking and financial world.

The problem facing the designer of such a system is this. The encryption process should disguise the message to such an extent that it is impossible to decode it without the decryption key. But since the essence of the system—indeed of any cipher system—is that the authorized receiver *can* decrypt the encoded message, the two keys must be mathematically related. Indeed, the receiver's program and decryption key *exactly undoes* the effect of the sender's program and encryption key, so it should

be theoretically possible to obtain the decryption key from the encryption key, provided one knows how the cipher programs work (and these days, anyone who wants to can find out).

The trick is to ensure that, although it is *theoretically* possible to recover the decryption key from the publicly available encryption key, it is *practically* impossible. In the case of the RSA system, the receiver's secret decryption key consists of a pair of large prime numbers (say, 75 digits each); the public encryption key consists of their product. Message encryption corresponds (very roughly) to multiplication of the two 75-digit primes, decryption corresponds (equally roughly) to factoring the 150-digit product, a task that is quite infeasible given present-day knowledge and technology. (The exact encryption and decryption procedures involve a generalization of Fermat's little theorem.)

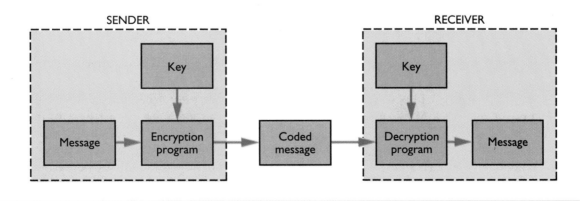

Numbers of the form M_n are nowadays known as *Mersenne numbers*. Calculation of the first few Mersenne numbers might lead one to suspect that M_n is prime whenever n is prime:

$$M_2 = 2^2 - 1 = 3$$
$$M_3 = 2^3 - 1 = 7$$
$$M_5 = 2^5 - 1 = 31$$
$$M_7 = 2^7 - 1 = 127,$$

all of which are prime. But then the pattern breaks down, with

$$M_{11} = 2,047 = 23 \times 89.$$

After that, the next Mersenne primes are M_{31}, M_{61}, M_{89}, M_{107}, and M_{127}.

What is the case is the opposite result, that M_n can be prime *only* when n is prime; all it takes to prove this assertion is a bit of elementary algebra. So, in looking for Mersenne primes, it is only necessary to look at Mersenne numbers M_n for which n is itself prime.

Another conjecture that many have been tempted to make on the basis of the numerical evidence is that M_n is prime whenever n is itself a Mersenne prime. The pattern holds until you reach the Mersenne prime $M_{13} = 8,191$; the 2,466-digit number $M_{8,191}$ is composite.

The task of finding Mersenne primes is made easier by virtue of a simple, reliable, and computationally efficient method to determine if a Mersenne number is prime. The method is based on Fermat's little theorem, but works only for Mersenne numbers. It is called the Lucas–Lehmer test. With the aid of this test, and many hours of time on some very powerful supercomputers, some thirty-three Mersenne primes have been discovered to date. The most recent is $M_{859,433}$, a prime number with over a quarter of a million digits, which was discovered early in 1994 on a Cray Y-MP M90 supercomputer in the United States. It only took about a half-hour to run the Lucas–Lehmer test on this one number, though many hours of computing time were spent checking other Mersenne numbers before the prime was found.

A Cray Y-MP M92 supercomputer, which has been used in the search for larger and larger prime numbers.

Why bother? Though there is generally a couple of inches of newspaper coverage for the discovery of the "world's largest known prime" (and, thanks to the efficiency of the Lucas–Lehmer test, the most recent world-record primes have all been Mersenne primes), the reason why endless hours of highly expensive supercomputer time are spent on this hunt is that searching for giant primes is an excellent way to test a new supercomputer system for accuracy.

Fermat's Last Theorem

Without a doubt, the most famous 'theorem' of mathematics is Fermat's last theorem. It is perhaps indicative of the huge gulf that separates mathematicians from most of humanity that of all the many remarkable results that have been proved over the years, the one 'theorem' of pure mathematics that is best known outside the field was not known to be a true theorem until 1994, despite a proof having been sought by some of the finest mathematicians for over three hundred years. The name 'last theorem' refers to the fact that, of all the results Fermat claimed to have proved, this was the last one to be decided one way or the other.

The story begins in 1670, five years after Fermat's death. His son, Samuel, was working his way through his father's notes and letters, with the intention of publishing them. Among these papers, there was a copy of Diophantus' book *Arithmetic*, the 1621 edition, edited by Claude Bachet, who had added a Latin translation alongside the original Greek text. It was this edition that had brought Diophantus' work to the attention of the European mathematicians, and it was clear from the various comments Fermat had written in the margins that the great French amateur had developed his own interest in number theory largely through the study of this master from the third century A.D.

Among Fermat's marginal comments were some forty-eight intriguing, and sometimes significant, observations, and Samuel decided to publish a new edition of *Arithmetic* with his father's notes appended. The second of these *Observations on Diophantus*, as Samuel called his father's comments, had been written by Fermat in the margin of Book II, alongside Problem 8.

Problem 8 asks: "Given a number which is a square, write it as a sum of two other squares." Written in algebraic notation, this problem says, given any number z, find two other numbers x and y such that

$$z^2 = x^2 + y^2.$$

Fermat's note read as follows:

> On the other hand, it is impossible for a cube to be written as a sum of two cubes or a fourth power to be written as a sum of two fourth powers or, in general, for any number which is a power greater than the second to be written as a sum of two like powers. I have a truly marvellous demonstration of this proposition which this margin is too narrow to contain.

Fermat's last theorem. This famous problem began as a comment Fermat wrote in the margin of his copy of Diophantus' classic text *Arithmetic*. After Fermat's death, his son published a new version *Arithmetic*, with his father's marginal comments added to the text.

Putting this text into modern notation, what Fermat claimed is that the equation

$$z^n = x^n + y^n$$

has no (whole-number) solutions for any power *n* greater than 2. (Mathematicians ignore the trivial solutions that arise when one of the unknowns is allowed to be zero.)

So began a saga that continued for over three centuries, as mathematician after mathematician, professional and amateur, attempted to produce a proof, perhaps the same proof Fermat had discovered, if such he had. In fact, it seems likely that Fermat was mistaken in his original belief, and subsequently realized his error. His marginal note was not, after all, intended for publication, so he would have had no reason to go back and erase it if he later found a flaw in his reasoning.

But for all that, the story was practically irresistible: a seventeenth-century amateur mathematician solves a problem that resists three hundred years of attack by the world's finest professional minds. Add the fact that the great majority of Fermat's claims did turn out to be correct, plus the simplicity of the statement itself, which any schoolchild can understand, and there is little reason to be surprised at the fame Fermat's last theorem has achieved. The offer of a number of large cash prizes for the first person to find a proof only added to the allure: in 1816, the French Academy offered a gold medal and a cash prize, and in 1908 the Royal Academy of Science in Göttingen offered another cash prize, the Wolfskell Prize.

It was its very resistance to proof that led to the theorem's fame. Fermat's last theorem has virtually no consequences, either in mathematics or in the everyday world. In making his marginal note, Fermat was simply observing that a particular numerical pattern that held for square powers did not hold for any higher powers; the interest was purely 'academic'. Had the issue been quickly decided one way or the other, the observation would have been worth nothing more than a footnote in subsequent textbooks.

And yet, had the problem been resolved early on, the mathematical world would likely have been a great deal the poorer. For the many attempts to solve the problem have led to the development of some mathematical notions and techniques whose importance for the rest of mathematics far outweighs that of Fermat's last theorem itself.

The problem in Diophantus' *Arithmetic* that started the whole affair (finding whole-number solutions to the equation $x^2 + y^2 = z^2$) is obviously related to the Pythagorean theorem. The question can be reformulated in an equivalent, geometric form as, do there exist right-angled triangles, all of whose sides are a whole number of units in length?

One well-known solution to this question is the triple $x = 3$, $y = 4$, $z = 5$:

$$3^2 + 4^2 = 5^2.$$

This identity was known as far back as the time of ancient Egypt, as illustrated in the box on the facing page. Is this the only solution? It does not take long to realize that the answer is no. Once you have found one solution, you immediately get a whole infinite family of solutions, since you can take your first solution and multiply all three values by any number you please, and the result will be another solution. Thus, from the 3, 4, 5 solution, you obtain the solution $x = 6$, $y = 8$, $z = 10$, the solution $x = 9$, $y = 12$, $z = 15$, and so on.

This trival means of producing new solutions from old can be eliminated by asking for (whole-number) solutions to the equation that have no factor in common to all three numbers. Such solutions are generally referred to as *primitive* solutions.

If only primitive solutions are allowed, are there any besides the 3, 4, 5 solution? Again, the answer is well known: the triple $x = 5$, $y = 12$, $z = 13$ is another primitive solution, and so too is the triple $x = 8$, $y = 15$, $z = 17$.

In fact, there are an infinite number of primitive solutions, and a complete resolution to the problem was given by Euclid in *Elements,* in the form of an exact pattern for all primitive solutions. The formulas

$$x = 2st, \qquad y = s^2 - t^2, \qquad z = s^2 + t^2$$

Right Angles in Egyptian Architecture

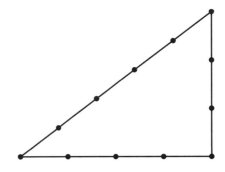

As long ago as 2,000 B.C., Egyptian architects knew that a triangle whose sides have lengths 3, 4, and 5 units is right angled, and used this knowledge for constructing right angles. They would first tie twelve equally long pieces of rope into a loop. Then, placing one of the knots at the point where they wanted to construct a right angle, they would pull the loop taut to form a triangle whose sides emanating from the starting point consisted of exactly three and four rope lengths, as shown. The resulting triangle would then be right angled, and so they would have their right angle.

This device is not an application of the Pythagorean theorem, but its converse: if a triangle is such that the sides are related by the equation

$$h^2 = a^2 + b^2,$$

then the angle opposite the side of length h is a right angle. This is Proposition I.48 in *Elements*. The Pythagorean theorem itself is Proposition I.47.

The Egyptian facility with right angles is illustrated by the square column, shown below, from the Temple of Luxor.

generate all the primitive solutions to the original equation, as s and t vary over all natural numbers such that:

(i) $s > t$;

(ii) s and t have no common factor;

(iii) one of s, t is even, the other odd.

Moreover, any primitive solution to the equation is of the above form for some values of s, t.

Turning now to Fermat's last theorem itself, there is some evidence to suggest that Fermat did have a valid proof in the case $n = 4$. That is to say, it is possible that he was able to prove that the equation

$$x^4 + y^4 = z^4$$

has no whole-number solutions. The evidence in question consists of one of the few complete proofs Fermat left behind: an ingenious argument to demonstrate that the area of a right-angled triangle whose sides are all whole numbers cannot be a square of a whole number. From this result, Fermat was able to deduce that the equation $x^4 + y^4 = z^4$ can have no whole-number solutions, and it is reasonable to assume that he established the result about triangles with square areas precisely in order to deduce this case of his 'last theorem'.

In order to establish his result about areas of triangles, Fermat's idea was to show that if there were natural numbers x, y, z such that $x^2 + y^2 = z^2$, and if, in addition, $\frac{1}{2}xy = u^2$ for some natural number u (i.e., if the area of the triangle is a square), then there are another four numbers x_1, y_1, z_1, u_1 that stand in the same relation to each other, for which $z_1 < z$.

Then, one may apply the same argument again to produce four more numbers, x_2, y_2, z_2, u_2, that also stand in the same relation, and for which $z_2 < z_1$.

But this process can go on forever. In particular, you will end up with an infinite sequence of nat-

Mathematical Induction

Mathematical induction is a powerful method of proof that provides the mathematician with a means to handle infinity. Suppose you have noticed some pattern, call it P, that seems to hold for every natural number n.

Perhaps when adding together more and more odd numbers, you notice that the sum of the first n odd numbers always seems to work out to be n^2:

$$
\begin{aligned}
1 + 3 &= 4 &&= 2^2 \\
1 + 3 + 5 &= 9 &&= 3^2 \\
1 + 3 + 5 + 7 &= 16 &&= 4^2 \\
1 + 3 + 5 + 7 + 9 &= 25 &&= 5^2 \\
1 + 3 + 5 + 7 + 9 + 11 &= 36 &&= 6^2
\end{aligned}
$$

and so on. You suspect that this pattern continues forever; that is to say, you suspect that *for every natural number n,* the following identity is true:

$$1 + 3 + 5 + \cdots + (2n - 1) = n^2.$$

Call this particular identity $P(n)$.

How do you *prove* that the identity $P(n)$ is true for every natural number n? The numerical evidence you have collected may seem pretty convincing—perhaps you have used a computer to verify that $P(n)$ holds for all n up to a billion. But numerical evidence alone can never provide you with a rigor-

ural numbers $z, z_1, z_2, z_3, \ldots$ such that

$$z > z_1 > z_2 > z_3 > \cdots .$$

But such an infinite sequence is impossible; eventually the sequence must descend to 1, and then it will stop.

ous proof, and indeed there have been a number of instances where numerical evidence of a billion or more cases has turned out *not* to be reliable. The problem is, the pattern P you are trying to verify is a pattern over the entire infinitude of all natural numbers. How can you show that a pattern holds for infinitely many objects? Certainly not by checking every single case.

This is where the method of mathematical induction comes in. In order to prove that a property $P(n)$ holds for *every* natural number n, it is enough to prove just two facts: first, $P(n)$ holds for the number $n = 1$ (i.e. $P(1)$ is true); second, if $P(n)$ is assumed to be true for an arbitrary number n, then $P(n + 1)$ follows. If you can establish both these facts, then you may conclude, without any further effort, that $P(n)$ is true for every natural number n.

The method of mathematical induction is extremely powerful, enabling you to draw a conclusion about a pattern holding for all the natural numbers, based on just two pieces of evidence. It is easy to appreciate the method in an intuitive way as a 'domino argument'. Suppose you were to stand a row of dominoes on their ends so that, if any one falls, it will knock over the next (if $P(n)$ is true for any n, then $P(n + 1)$ will hold). Then, if you knock over the first domino (if $P(1)$ holds), the entire row of dominoes will fall down ($P(n)$ holds for every n), as each domino is knocked over by its predecessor and in turn knocks down the next one in the row.

Using the method of induction, it is straightforward to verify the particular pattern P given above. For $n = 1$, the identity just says

$$1 = 1^2,$$

which is trivial. Now assume the identity holds for an arbitrary number n. That is, assume that

$$1 + 3 + 5 + \cdots + (2n - 1) = n^2.$$

The next odd number after $(2n - 1)$ would be $(2n + 1)$; add this number to both sides of this identity, to give

$$1 + 3 + 5 + \cdots + (2n - 1) + (2n + 1) \\ = n^2 + (2n + 1).$$

By elementary algebra, the expression following the equals sign reduces to $(n + 1)^2$. Hence you can rewrite the last identity as

$$1 + 3 + 5 + \cdots + (2n - 1) + (2n + 1) \\ = (n + 1)^2.$$

This is just the identity $P(n + 1)$. So, the above argument shows that, *if* $P(n)$ holds for some n, *then* $P(n + 1)$ follows. Thus, by mathematical induction, you may conclude that $P(n)$ does indeed hold for every natural number n.

Hence there can be no numbers x, y, z with the supposed properties, which is what Fermat set out to prove.

For obvious reasons, this method of proof is referred to as Fermat's *method of infinite descent*. It is closely related to the present-day *method of mathematical induction*, a powerful tool for verifying many patterns involving natural numbers. Mathematical induction is further explained in the box on this page.

With the case of exponent $n = 4$ established, mathematicians readily observed that if Fermat's last theorem is also true for all prime exponents, then it must be true for all exponents. So, the would-be 'last

Leonhard Euler (1707–1783).

theorem prover' was faced with the case of an arbitrary prime exponent.

The first person to make any real progress in this direction was Euler. In 1753, he claimed to have proved the result for $n = 3$. Though his published proof contained a fundamental flaw, the result is generally still credited to him. The problem with Euler's proof was that it depended upon a particular assumption about factorization that Euler made in the course of his argument. Though this assumption can in fact be proved for the case $n = 3$, it is not true for all prime exponents, as Euler seemed to be assuming, and in fact it was precisely this subtle, but invalid, assumption that brought down many subsequent attempts to prove Fermat's last theorem.

In 1825, extending Euler's argument, Peter Gustav Lejeune Dirichlet and Adrien-Marie Le-gendre proved Fermat's last theorem for exponent $n = 5$. (Their version of the argument avoided the factorization trap that befell Euler.)

Then, in 1839, using the same general approach, Gabriel Lamé proved the result for $n = 7$. By this stage, the argument was becoming increasingly intricate, and there seemed little hope of taking it any further, to deal with the next case, $n = 11$. (Not that this kind of piecemeal approach would solve the entire problem in any case.)

To make any further progress, what was required was the detection of some kind of general pattern in the proofs, a way of stepping back from the complexity of the individual trees to the larger-scale order of the forest. This advance was made by the German mathematician Ernst Kummer in 1847.

Kummer recognized that some prime numbers exhibited a certain kind of pattern, referred to by Kummer as *regularity*, which enabled an Euler-type proof of Fermat's last theorem to be carried through. Using this new property of regularity, Kummer was able to prove that Fermat's last theorem holds for all exponents n that are regular primes. Of the primes less than 100, only 37, 59, and 67 fail to be regular, so in one fell swoop Kummer's result established Fermat's last theorem for all exponents up to 36 and for all prime exponents less than 100 apart from 37, 59, and 67.

There are a number of different, but totally equivalent ways to define exactly what a regular prime is, but all refer to some fairly advanced mathematical concepts, so I will not give any definition here. What I will tell you is that computer searches as far as 4,000,000 showed that most primes were regular.

Moreover, all the nonregular primes less than 4,000,000 were shown to satisfy a property a bit weaker than regularity, but which still implies Fermat's last theorem for that exponent. So the Kummer approach, together with computational work, enabled mathematicians to conclude that Fermat's last theorem was known to be true for all exponents up to 4,000,000. But was it true for *all* exponents, as Fermat had claimed?

At this point, we must leave Fermat's last theorem for the time being. We shall come back to it in Chapter 6, when I shall tell you about a startling discovery made in 1983, and the even more unexpected and highly dramatic end to the Fermat saga that took place in 1993 and 1994. The reason for putting off these two developments until later—

indeed, several chapters later—is itself a striking illustration that mathematics is the search for, and study of, patterns. The 1983 and 1993–94 discoveries only came about as a result of investigations of patterns of quite different natures—not number patterns but patterns of shape and position, patterns that involve the infinite in a fundamental way.

Nicolaus Neufchatel, *Portrait of Johannes Neudorfer and his Son* (1561).

2

Reasoning and Communicating

O n page 17, there is a 'proof' of the fact that the number $\sqrt{2}$ is not rational, that is to say, cannot be expressed as a ratio of two whole numbers. Anyone who follows the discussion carefully, and thinks about each step, will surely find it entirely convincing—it does, indeed, *prove* the assertion that $\sqrt{2}$ is irrational. But just what is it about that particular discussion, about those particular English sentences written in that particular order, that makes it a proof?

Admittedly, the argument does make use of some simple algebraic notation; but that is not the crux of the matter. It would be extremely easy to eliminate all the algebra, replacing every symbol by words and phrases of the English language, and the result would still be a proof of the assertion—indeed, it would be *the same proof!* The choice of language, whether symbolic, verbal, or even pictorial, might affect the length of the proof, or the ease with which you can understand it, but it does not affect whether the argument does or does not constitute a proof. In human terms, being a proof means having the capacity to completely convince *any* sufficiently educated, intelligent, rational person, and surely that has to do with some kind of *abstract pattern* or *abstract structure* associated with the argument. What is that abstract structure, and what can be said about it?

Even more fundamental, the same question can be posed about language itself. Proofs are just one of many things that can be expressed by language. What is it about the symbols on this page that enables me, as author, to communicate my thoughts to you, as reader? As in the case of proofs, *the same thoughts* will be communicated by any foreign-language edition of this book, so again the answer is surely not something physical and concrete about this particular page or the particular language used, but has to be some kind of *abstract structure* associated with what appears on the page. What is this abstract structure?

The abstract patterns that human beings are equipped to recognize and utilize are not to be found solely in the physical world; there are abstract patterns involved in our thinking and communicating with one another.

Aristotle (ca. 340 B.C.).

Greek Logic

The first systematic attempt to describe the patterns involved in proof was made by the ancient Greeks, in particular Aristotle. These efforts resulted in the creation of *Aristotelian logic*. (It is not clear exactly how much of this work is due to Aristotle himself, and how much to his followers. Whenever I use the term 'Aristotle', I am referring to Aristotle and his followers.)

According to Aristotle, a proof, or 'rational' argument, or 'logical' argument, consists of a series of assertions, each one following 'logically' from previous ones in the series, according to some 'logical rules'. Of course, this definition can't be quite right, since it does not provide any means for the proof to begin: the first assertion in an argument cannot follow from any previous assertions, since in this case there are no previous assertions! But any proof must depend on some initial facts or assumptions, so the series can start off by listing those initial assumptions, or at least some of them. (In practice, the initial assumptions may be 'obvious' or 'understood', and might, therefore, not be mentioned explicitly. Here I shall follow normal mathematical practice and concentrate on the ideal case, where all steps are present.)

Aristotle's next step was to describe the logical rules that may be used to arrive at a valid conclusion. To handle this issue, he assumed that any correct argument may be formulated as a series of assertions of a particular form: the so-called subject–predicate proposition.

By a *proposition* is simply meant a sentence that is either true or false. The *subject–predicate* propositions considered by Aristotle are those consisting of two entities, a subject and a property, or predicate, ascribed to that subject. Examples of such propositions are:

Aristotle is a man.

All men are mortal.

Some musicians like mathematics.

No pigs can fly.

You might well wonder if Aristotle was correct in assuming that any valid argument could be broken down into a series of assertions of this particularly simple form. The answer is that this is not at all the case; for instance, many mathematical proofs cannot be analyzed in this manner. Even in cases where such an analysis is possible, it can be extremely difficult to actually break the argument down into steps of this kind. So Aristotle's analysis did not, in fact, identify an abstract structure applicable to *all* correct arguments; rather, his analysis applies only to a certain, very restricted kind of correct argument.

What makes Aristotle's work nevertheless of lasting historical value is that not only did he look for patterns in correct argument, he actually did find some. It was to be almost two thousand years before anyone took the study of patterns of rational argument significantly further!

The logical rules that Aristotle identified as the patterns that must be followed in order to construct a correct proof (using subject–predicate propositions) are known as *syllogisms*. These are rules for deducing one assertion from exactly two others.
An example of a syllogism is:

All men are mortal.

Socrates is a man.

Socrates is mortal.

The idea is that the third assertion, the one below the line, follows 'logically' from the previous two. In the case of this simple—and very hackneyed—example, this deduction certainly seems correct enough, albeit pretty obvious. What makes Aristotle's contribution so significant is that he abstracted a general pattern from such examples.

His first step was to abstract away from any particular example to obtain a general case. Let S denote the subject of any subject–predicate proposition, P the predicate. In the case of the proposition *Socrates is a man*, S denotes Socrates and P denotes the predicate 'is a man'. This step is very much like the step of replacing numbers by algebraic symbols such as the letters x, y, and z. But, instead of the symbols S and P denoting arbitrary *numbers*, they denote an arbitrary *subject* and an arbitrary *predicate*. Eliminating the particular in this way sets the stage to examine the abstract patterns of reasoning.

According to Aristotle, the predicate may be used affirmatively or negatively in the proposition, like this:

S *is* P or S *is not* P.

Moreover, the subject may be *quantified*, by expressing it in the form *all S* or *some S*.

The two kinds of quantifications of the subject may be combined with the two possibilities of affirmative and negative predicates to give a total of four possible quantified subject–predicate propositions:

All S is P.

All S is not P.

Some S is P.

Some S is not P.

The second of these is easier to read—and more grammatical—if it is rewritten in the equivalent form *No S is P*.

It might also seem more grammatical to write the first of these four in the plural, as *All S are P*.

But this is a minor issue that disappears with the next step in the abstraction process, which is to write the four syllogism patterns in abbreviated form:

SaP: All S is P.

SeP: No S is P.

SiP: Some S is P.

SoP: Some S is not P.

These abbreviated forms make very clear what are the abstract patterns of propositions that Aristotle was looking at.

Most work with syllogisms concentrates on propositions of the four quantified forms just stated. On the face of it, these forms would appear to ignore examples like the one given earlier, *Socrates is a man*. But examples such as this, where the subject is a single individual, are indeed still covered. In fact, they are covered twice. If S denotes the collection of all 'Socrateses', and P denotes the predicate of being a man, then either of the forms SaP and SiP captures this particular proposition. The point is that, because there is only one Socrates, all of the following are equivalent:

'Socrates',   'all Socrateses',   'some Socrates'.

In everyday English, only the first of these seems sensible. But the whole purpose of this process of abstraction is to get away from ordinary language and work with the abstract patterns expressed in the language.

The decision to ignore individual subjects, and concentrate instead on collections or kinds of subject, has a further consequence: the subject and predicate in a subject–predicate proposition may be interchanged. For example, *All men are mortal* may be changed to *All mortals are men*. Of course, the changed version will generally mean something quite different from the original, and may be false or even nonsensical; but it still has the same abstract structure as the original, namely, *all 'somethings' are 'somethings'*. In regard to allowable subject–predicate propositions, the four construction rules listed a mo-

ment ago are *commutative*: the terms S and P may be interchanged in each one.

Having described the abstract structure of the propositions that may be used in an Aristotelian argument, the next step is to analyze the syllogisms that may be constructed using those propositions. What are the valid rules that may be used in order to construct a correct argument as a series of syllogisms?

A syllogism consists of two initial propositions, called the premises of the syllogism, and a conclusion that, according to the rule, follows from the two premises. If S and P are used to denote the subject and predicate of the conclusion, then, in order for inference to take place, there must be some third entity involved in the two premises. This additional entity is called the middle term; denote it by M.

For the example

All men are mortal.

Socrates is a man.
_____

Socrates is mortal.

S denotes Socrates, P denotes the predicate of being mortal, and M is the predicate of being a man. In symbols, this particular syllogism thus has the form

$$MaP$$
$$SaM$$
_____
$$SaP$$

(The second premise and the conclusion could also be written using *i* instead of *a*.) The premise involving M and P is called the major premise, and is written first; the other one, involving S and M, is called the minor premise, and is written second.

Having standardized the way syllogisms are represented, an obvious question to ask is, how many possible syllogisms are there?

Each major premise can be written in one of two orders, with the M first or the P first. Similarly, there are two possible orders for the minor premise, S first or M first. The possible syllogisms thus fall

Venn Diagrams

The nineteenth-century English mathematician John Venn invented a simple, geometric method for checking the validity of syllogisms, known nowadays as the method of Venn diagrams. In Venn's method, the syllogism is represented by means of three overlapping circles, as shown. The idea is that the region inside the circle marked S represents all objects of type S, and analogously for circles marked P and M. The procedure used to verify the syllogism is to see what the two premises say about various regions in the diagram, numbered 1 to 7.

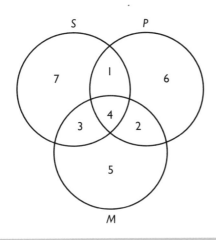

To illustrate the method, consider the simple example given in the text, namely, the syllogism

$$MaP$$
$$\underline{SaM}$$
$$SaP$$

The major premise, *All M are P*, says that regions 3 and 5 are empty. (All the objects in M are in P, that is to say, are in the regions marked 2 and 4.) The minor premise, *All S are M*, says that regions 1 and 7 are empty. Thus, the combined effect of the two premises is to say that regions 1, 3, 5, and 7 are empty.

The goal now is to construct a proposition involving S and P that is consistent with this information about the various regions. Regions 1, 3, and 7 are empty, so anything in S must be in region 4, and is therefore in P. In other words, all S are P. And that verifies this particular syllogism.

Though not all syllogisms are as easy to analyze as this example, all the other valid syllogisms can be verified in the same way.

into four distinct classes, known as the *figures* of the syllogism:

I	II	III	IV
MP	*PM*	*MP*	*PM*
SM	*SM*	*MS*	*MS*
SP	*SP*	*SP*	*SP*

For each figure, the gap between the subject and the predicate of each proposition can be filled with any one of the four letters *a*, *e*, *i*, or *o*. So that gives a total of 4 × 4 × 4 × 4 = 256 possible syllogisms.

Of course, not all possible patterns will be logically valid, and one of Aristotle's major accomplishments was to find all the valid ones. Of the 256

possible syllogism patterns, Aristotle's list of the valid ones consisted of precisely the nineteen listed below. (Aristotle made two errors, however. His list contains two entries that do not correspond to valid inferences, as the next section will show.)

First figure: *aaa, eae, aii, eio*

Second figure: *eae, aee, eio, aoo*

Third figure: *aai, iai, aii, eao, oao, eio*

Fourth figure: *aai, aee, iai, eao, eio*

The box shown above explains an elegant, and more recent, way to check the validity of syllogisms, invented by Venn. Using simple geometric ideas, it is called the method of Venn diagrams.

For all its simplicity, the method of Venn diagrams is remarkable, in that it provides a *geometric* way to think about deduction. In the above discussion, patterns of thought were transformed first to algebraic patterns and then to simple geometric patterns. This demonstrates yet again the incredible power of the mathematical method of abstraction.

Having narrowed down the collection of possible syllogisms to the valid ones, further simplifications can be made to Aristotle's list, by removing any syllogism whose logical pattern is duplicated elsewhere in the list. For instance, in any proposition involving an *e* or an *i*, the subject and predicate may be interchanged without affecting the meaning of the proposition, and each such interchange will lead to a logical duplicate. When all of the various redundant patterns have been removed, only the following eight forms are left:

I   *aaa, eae, aii, eio*

II   *aoo*

III   *aai, eao, oao*

The fourth figure has disappeared altogether.

The two invalid syllogisms in Aristotle's list, mentioned earlier, are still there. Now can you spot them? Try using the method of Venn diagrams to check each one. If you fail to find the mistakes, take heart. It took almost two thousand years for the errors to come to light. The next section explains how the mistake was corrected. The length of time it took to resolve the issue was not due to any lack of attention to the syllogism. Indeed, over the centuries, Aristotle's logical theory achieved an exalted status in human learning. For instance, as recently as the fourteenth century, the statutes of the University of Oxford included the rule "Bachelors and Masters of Arts who do not follow Aristotle's philosophy are subject to a fine of five shillings for each point of divergence." So there!

Boole's Logic

From the time of the ancient Greeks until the nineteenth century virtually no advances were made in the mathematical study of the patterns of rational argument. The first breakthrough since Aristotle came with the arrival on the mathematical scene of the Englishman George Boole.

Born in 1815 to Irish parents in East Anglia, Boole grew to mathematical maturity at a time when mathematicians were starting to realize that algebraic symbols could be used to denote entities other than numbers, and that the methods of algebra could be applied to domains other than ordinary arithmetic. For instance, the end of the eighteenth century had seen the development of the arithmetic of complex numbers, a generalization of the real numbers that we will encounter in Chapter 3. There was also Hermann Grassmann's development of vector algebra. (A vector is an entity that has both magnitude and direction, such as velocity or force. Vectors can be studied both geometrically and algebraically, as discussed in the box on page 44.)

Boole set about trying to capture the patterns of thought in an algebraic fashion. In particular, he sought an algebraic treatment of Aristotle's syllogistic logic. Of course, it is an easy matter simply to *represent* Aristotle's own analysis using algebraic symbols, as I did in the previous section. But Boole went much further; he provided an algebraic treatment of the logic, using not only algebraic notation but algebraic structure as well.

Boole's brilliant analysis was described in a book he published in 1854, titled *An Investigation of the Laws of Thought on Which are Founded the Mathematical Theories of Logic and Probabilities*. Often referred to more simply as *The Laws of Thought*, this pivotal book built upon an earlier, and much less well known, treatise of Boole's, titled *The Mathematical Analysis of Logic*.

Chapter 1 of *The Laws of Thought* begins with these words:

The design of the following treatise is to investigate the fundamental laws of those operations of the mind by which reasoning is performed; to give expression to them in the symbolic language of a Calculus, and upon this foundation to establish the science of Logic and construct its method.

The starting point of Boole's logic is the same idea that led to the method of Venn diagrams, namely, to regard propositions as dealing with *classes* or *collections* of objects, and to reason with those collections. For example, the proposition *All men are mortal* can be taken to mean that the class of all men is a subclass (or subcollection, or part) of the class of all mortals. Another way to say this is, the members of the class of all men are all members of the class of all mortals. But Boole did not look for structure at the level of the members of these classes; rather, he concentrated on the classes themselves, developing an 'arithmetic' *of classes*. The idea is both simple and elegant, and was to prove extremely powerful.

George Boole (1815–1864).

Start off by using letters to denote arbitrary collections of objects, say x, y, z. Denote by xy the collection of objects common to both x and y, and write $x + y$ to denote the collection of objects that are in either x or y or in both. (Actually, in defining the 'addition' operation, Boole distinguished the case where x and y have no members in common from the case where they overlap. Modern treatments—and this book is one such—generally do not make this distinction.)

Let 0 denote the empty collection, and let 1 denote the collection of all objects. Thus, the equation $x = 0$ means that x has no members. The collection of all objects not in x is denoted by $1 - x$.

Boole observed that his new 'arithmetic' of collections has the following properties:

$$x + y = y + x \qquad xy = yx$$
$$x + (y + z) = (x + y) + z \qquad x(yz) = (xy)z$$
$$x(y + z) = xy + xz$$
$$x + 0 = x \qquad 1x = x$$
$$2x = x + x = x \qquad x^2 = xx = x$$

The first five of these identities are familiar properties of ordinary arithmetic, where the letters denote numbers; they are the two commutative laws, the two associative laws, and the distributive law.

The next two identities indicate that 0 is an identity operation for 'addition' (that is, Boole's 0 behaves like the number 0) and 1 is an identity operation for 'multiplication' (that is, Boole's 1 behaves like the number 1).

The last two identities appear quite strange when you see them for the first time; they certainly are not true for ordinary arithmetic. They are called the idempotent laws.

In modern terminology, any collection of objects and any two operations ('multiplication' and 'addition') on them that obey all of the above identities is called a *Boolean algebra*. In fact, the system as just described is not exactly the one Boole himself developed; in particular, his treatment of

Vector Algebra

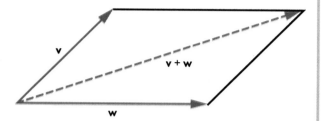

A *vector* is an abstract entity that has magnitude and direction (in the plane, in three-dimensional space, or in a space of four or more dimensions). Examples of vectors are velocity, acceleration, and force. From these three examples alone, it is clear that vectors arise with considerable frequency in physics, and indeed it was physics that provided the motivation for the development of an 'algebra' of vectors. Credit for the introduction of *vector algebra* is normally given to the German mathematician Hermann Grassmann, who described the new theory in his book *Ausdehnungslehre*, published in 1844.

There are a number of 'arithmetic' operations that can be performed on a vector. For instance, since a vector has magnitude, you can 'multiply' it by any number: if the vector **v** is 'multiplied' by the number k, the resulting vector, k**v**, has the same direction as **v** and magnitude k times the magnitude of **v**. For example, if **v** denotes the velocity of a woman driving due north at 30 m.p.h., then 2**v** denotes her velocity when she is driving due north at 60 m.p.h. (If k is negative, the vector changes direction as well. For example, multiplying the vector for driving due north at 30 m.p.h. by −1.5 gives the vector for driving due south at 45 m.p.h.)

Another 'arithmetic' operation for vectors is 'addition'. In physics, the vector sum corresponds to the resultant force when two forces are applied, such as the force on an airplane subjected to a lateral wind. In the case of vectors in the plane, the simplest way to understand the addition of two vectors is to use the so-called parallelogram law. If **v** and **w** are two vectors, you can obtain the vector sum **v** + **w** by drawing the arrows for **v** and **w** so that they both start at the same point, and then completing the parallelogram, as shown. The diagonal that starts at the same point as **v** and **w** represents **v** + **w**.

The 'arithmetic' operations of multiplication of a vector by a number and addition of two vectors have the properties listed below. To distinguish between vectors and numbers, it is common to denote vectors by boldface type. The 'zero vector', **0**, has no magnitude and no direction.

$$\mathbf{v} + \mathbf{w} = \mathbf{w} + \mathbf{v}$$

$$\mathbf{u} + (\mathbf{v} + \mathbf{w}) = (\mathbf{u} + \mathbf{v}) + \mathbf{w}$$

$$a(b\mathbf{v}) = (ab)\mathbf{v}$$

$$c(\mathbf{v} + \mathbf{w}) = c\mathbf{v} + c\mathbf{w}$$

$$(c + d)\mathbf{v} = c\mathbf{v} + d\mathbf{v}$$

$$\mathbf{v} + \mathbf{0} = \mathbf{v}$$

$$1\mathbf{v} = \mathbf{v}$$

Any collection of entities having the above properties is called a *vector space*. More precisely, if you have a collections of entities called vectors, an operation of addition of two vectors to give a third vector, and an operation of multiplication of a vector by a number to give another vector, and if these operations have all of the properties listed above, then the entire system is called a vector space. The study of such systems forms an extremely important branch of mathematics known as linear algebra.

It is interesting—and extremely useful—that the above axioms for a vector space make no mention of magnitude or direction. Thus, the concept of a vector space may be found in areas other than physics, the area for which the notion was originally developed. In particular, linear algebra has applications in many walks of life, and is taught to all college students of mathematics, physics, engineering, computer science, economics, and business.

'addition' meant that his system did not have the idempotent law for addition. As is often the case in mathematics—and, indeed, in any walk of life—there is always room for others to improve upon a good idea; in this case, Boole's system was subsequently modified to give the one described here.

Boole's algebraic logic provides an elegant way to study Aristotle's syllogisms. In Boole's system, the four kinds of subject–predicate proposition considered by Aristotle can be expressed like this:

$$SaP: \quad s(1 - p) = 0$$

$$SeP: \quad sp = 0$$

$$SiP: \quad sp \neq 0$$

$$SoP: \quad s(1 - p) \neq 0$$

With the syllogisms represented this way, it is a matter of simple algebra to determine which are valid. For example, take a syllogism in which the two premises are:

All P are M.

No M is S.

Expressing these algebraically, you get the two equations

$$p(1 - m) = 0,$$

$$ms = 0.$$

By ordinary algebra, the first of these can be rewritten like this:

$$p = pm.$$

Then, by playing around for a few moments to find an expression involving just p and s (the conclusion must not involve the middle term), you discover that

$$ps = (pm)s = p(ms) = p0 = 0.$$

In words:

No P is S.

And that is really all there is to using Boole's logic. Everything has been reduced to elementary algebra, except that the symbols stand for propositions rather than numbers. (Just as Grassmann was able to reduce arguments about vectors to algebra.)

It was by using Boole's algebraic logic that two mistakes were found in Aristotle's original treatment of the syllogism. Two of the forms listed by Aristotle as valid are not, in fact, valid. They are both in the third figure: *aai* and *eao*.

In words, the first of these forms says:

All M are P.

All M are S.

Some S is P.

Written algebraically, it looks like this:

$$m(1 - p) = 0$$

$$m(1 - s) = 0$$

$$sp \neq 0$$

The question then becomes, does the third equation follow from the first two? The answer is that it does not. If $m = 0$, then the first two equations are true, *whatever s and p denote*. Consequently, it is possible for the two premises to be true when the conclusion is false, so the form is invalid. For example, according to one invalid form, the two valid premises

All green pigs are green.

All green pigs are pigs.

would yield the (invalid) conclusion

Some pigs are green.

The same kind of thing happens with the other syllogism that Aristotle misclassified.

It is true that, if $m \neq 0$, both syllogisms work out fine. And this is undoubtedly why the error was not spotted for well over a thousand years. When you are thinking in terms of words, about *predicates*, it is not natural to ask what happens if one of the predicates describes something that is impossible. But when you are manipulating simple algebraic equations, it is not only natural but, for the mathematician, second nature to check whether terms are zero or not.

Translating patterns of logic to patterns in algebra does not change the patterns in an intrinsic way. But it does change the way people can think about those patterns. What is unnatural and difficult in one framework can become natural and easy in another. In mathematics as in other walks of life, it is often not just *what* you say that matters, but the *way* that you say it.

Propositional Logic

Boole's algebraic system succeeded, and succeeded extremely well, in capturing Aristotle's syllogistic logic. But its importance rests on far more than that one achievement.

For all its intrinsic interest, Aristotle's system is simply too narrow in its scope. Though many arguments can be recast as a series of subject–predicate propositions, this is not always the most natural way to express an argument; and besides, many arguments simply cannot be made to fit the syllogistic mold.

Starting with Boole's algebraic treatment of reasoning, logicians took a much more general approach to finding the patterns used in deduction. Instead of studying arguments involving propositions of a particular kind, as Aristotle had done, they allowed any proposition whatsoever.

When using this approach, you start off with some 'basic', unanalyzed propositions. The only thing you know about these propositions is that they *are* propositions; that is, they are statements that are either true or false (though, in general, you do not know which). A number of precisely stipulated rules (outlined below) enable you to combine these basic

propositions to produce more complex propositions. You then analyze those arguments that consist of a series of such compound propositions.

This system is known as *propositional logic*. It is highly abstract, since the logical patterns uncovered are entirely devoid of any content. The theory is completely independent of what the various propositions say.

Like Aristotle's syllogistic logic, propositional logic still has the defect of being too restrictive; not all arguments are of this kind. Nevertheless, a great many arguments can be so analyzed. Moreover, the 'logical patterns' that are uncovered by this approach do provide considerable insight into the notion of mathematical proof and, indeed, of logical deduction in general. Since the theory is independent of what the various propositions are about, the patterns uncovered are those of *pure logic*.

Most of the rules for combining propositions to give more complex, compound propositions are essentially those considered by Boole in his algebraic treatment. However, the description I shall give here is a result of subsequent refinements by a number of other logicians. (What is more, my order of presentation does not accord with the historical development.)

Since the only fact known about a proposition is that it is true or false, it is hardly surprising that the notion of truth and falsity plays a central role in this theory. The logical patterns that arise when propositions are combined are *patterns of truth*.

For instance, one way to combine propositions is by the operation of *conjunction*: given propositions *p* and *q*, form the new proposition *p and q*. For example, the conjunction of the two propositions *John likes ice cream* and *Mary likes pineapple* is the compound proposition *John likes ice cream and Mary likes pineapple*. In general, all you can hope to know about the compound proposition [*p and q*] is what its truth status is, given the truth status of *p* and *q*. A few moments reflection gives the appropriate pattern. If both *p* and *q* are true, then the conjunction [*p and q*] will be true; if one or both of *p* and *q* are false, then [*p and q*] will be false.

The pattern is perhaps most clearly presented in tabular form, as a 'truth table'. The box on the following page gives the truth table for conjunction, and for three other operations on propositions, namely, *disjunction* [*p or q*], the *conditional* [*p → q*], and *negation* [*not p*].

The last of these, [*not p*], is self-explanatory, but the other two require some comment. In everyday language, the word 'or' has two meanings. It can be used in an exclusive way, as in the sentence "The door is locked *or* it is not locked." In this case, only one of these possibilities can be true. Alternatively, it may be used inclusively, as in "It will rain *or* snow." In this case, there is the possibility that both will occur. In everyday communication, people generally rely on context to make the intended meaning clear. But in the logic of propositions, there is no context, only the bare knowledge of truth or falsity. Since mathematics requires unambiguous definitions, mathematicians had to make a choice when they were formulating the rules of propositional logic. They chose the inclusive version. Their decision is reflected in the truth table for disjunction.

Since it is an easy matter to express the exclusive-or in terms of the inclusive-or and the other logical operations, there is no net loss in making this particular choice. But the choice was not arbitrary. Mathematicians chose the inclusive-or because it leads to a logical pattern much more similar to that of Boolean algebra, described in the previous section.

There is no common English word that directly corresponds to the conditional. It is related to logical implication, so "implies" would be the closest word. But the conditional does not really capture the notion of implication. Implication involves some kind of *causality*; if I say that *p implies q* (an alternative way to say the same thing is *If p, then q*), you will understand there to be some sort of connection between *p* and *q*. But the operations of propositional logic are to be defined solely in terms of truth and falsity, and this method of definition is too narrow to capture the notion of implication. The conditional operation does its best, by capturing the two patterns of truth that arise from an implication, namely:

(i) if it is the case that *p* implies *q*, then the truth of *q* follows from that of *p*, and

(ii) if it is the case that *p* is true and *q* is false, then it cannot be the case that *p* implies *q*.

These considerations give you the first two rows of the truth table for the conditional. The remainder of the truth table, which concerns the two cases when *p* is false, is completed in a fashion that leads to the most useful theory. Here you have an example of being guided by a *mathematical* pattern, when there is no real-world pattern to go by.

Patterns of truth explain the rules for combining propositions. But what are the patterns involved in deducing one proposition from another? Specifically, in propositional logic, what takes the place of Aristotle's syllogisms?

The answer is the following, simple deduction rule, known as *modus ponens*:

From p → q and p, infer q.

This rule clearly accords with the intuition that the conditional corresponds to the notion of implication.

It should be stressed that the *p* and *q* here do not have to be simple, noncompound propositions. As far as modus ponens is concerned, these symbols may denote any proposition whatsoever. Indeed, throughout propositional logic, the algebraic symbols used almost invariably denote arbitrary propositions, simple or compound.

In propositional logic, a *proof*, or *valid deduction*, consists of a series of propositions such that each proposition in the series is either deduced from previous ones by means of modus ponens, or else is one of the assumptions that underlie the proof. During the course of a proof, any of the logical identities listed in the box on the next page may be used, just as any of the laws of arithmetic may be used in the course of a calculation.

Patterns of Truth

The logical operations used to combine propositions into more complex propositions are formally defined by showing how the truth or falsity of the compound is related to that of the individual components. This type of definition is most conveniently represented in a tabular form, as a so-called truth table.

The tables below, which capture certain 'patterns of truth', are labeled using the mathematical symbols commonly used nowadays: $\wedge$ to denote conjunction (*and*), $\vee$ to denote disjunction (*or*), and $\neg$ to denote negation (*not*). These tables provide the formal definitions of these logical operations. This use of truth tables motivates the use of the term 'truth *value*' to denote the truth (value = T) or falsity (value = F) of a particular proposition.

Reading along a row, each table indicates the truth value of the compound that arises from the truth values of the components.

p q	$p \wedge q$	p q	$p \vee q$	p q	$p \to q$
T T	T	T T	T	T T	T
T F	F	T F	T	T F	F
F T	F	F T	T	F T	T
F F	F	F F	F	F F	T

p	$\neg p$
T	F
F	T

Since the various logical operations are defined purely in terms of their truth patterns, if two compound propositions have truth tables that are row-by-row identical, then the two compounds are, to all intents and purposes, equal. By computing truth tables, the following 'laws of logical algebra' can be obtained:

$$p \wedge q = q \wedge p \qquad p \vee q = q \vee p$$

$$p \wedge (q \wedge r) = (p \wedge q) \wedge r$$

$$p \vee (q \vee r) = (p \vee q) \vee r$$

$$p \wedge (q \vee r) = (p \wedge q) \vee (p \wedge r)$$

$$p \vee (q \wedge r) = (p \vee q) \wedge (p \vee r)$$

$$p \wedge T = p \qquad p \vee T = T$$

$$p \wedge F = F \qquad p \vee F = p$$

$$\neg(p \wedge q) = (\neg p) \vee (\neg q)$$

$$\neg(p \vee q) = (\neg p) \wedge (\neg q)$$

$$\neg\neg p = p$$

$$p \to q = (\neg p) \vee q$$

In these equations, T denotes any true proposition, such as $1 = 1$, and F denotes any false proposition, such as $0 = 1$.

Apart from the last one, these equations exhibit some similarities to the familiar laws of arithmetic. A much stronger connection unites the above patterns and those in Boole's logic. If $p \wedge q$ is taken to correspond to Boole's pq, $p \vee q$ to Boole's $p + q$, and $\neg p$ to Boole's $1 - p$, and if T, F are taken to correspond to Boole's 1, 0, respectively, then all of the above identities (apart from the last one) hold for Boole's logic.

In all of these identities, the 'equality' is not genuine equality. All it means is that the two propositions concerned have the same truth table. In particular, with this meaning of 'equals' the following is true:

7 is a prime = *the angles of a triangle sum to* $180°$.

Because of this very special meaning of 'equality', mathematicians generally use a different symbol, writing $\equiv$ or $\leftrightarrow$ instead of $=$ in such identities.

Though it does not capture all kinds of reasoning, not even all kinds of mathematical proof, propositional logic has proved to be extremely useful. In particular, to all intents and purposes, today's electronic computer is simply a device that can perform deductions in propositional logic. Indeed, the two great pioneers of computing, Alan Turing and John von Neumann, were both specialists in mathematical logic.

Predicate Logic and Patterns of Language

The final step in trying to capture the patterns involved in mathematical proofs was supplied by Guiseppe Peano and Gottlob Frege at the end of the nineteenth century. Their idea was to take propositional logic and add further deductive mechanisms that depend on the nature of the propositions, not just their truth values. In a sense, their system of logic combined the strengths of both Aristotle's approach—where the rules of deduction do depend on the nature of the propositions—and propositional logic, which captures deductive patterns of 'pure logic'. However, the additional rules used to produce *predicate logic*, as the new theory is known, are

A microprocessor. The heart of today's electronic computer, the so-called silicon chip is essentially a complex array of patterns of propositional logic, etched into silicon.

far more general than Aristotle's syllogism. (Though they do involve the notions of 'all' and 'some' that formed the basis of the figures of the syllogism.)

Predicate logic has no unanalyzed, 'atomic' propositions. All propositions are regarded as built up from more basic elements. In other words, in predicate logic, a study of the patterns of deduction is preceded by, and depends upon, a study of certain *linguistic* patterns—the patterns of language used to form propositions.

This system of logic takes as its basic elements not propositions but *properties*, or *predicates*. The more simple of these are the very same constituents of Aristotle's logic, predicates such as:

> *. . . is a man.*

> *. . . is mortal.*

> *. . . is an Aristotle.*

However, predicate logic allows for more complex predicates, involving two or more objects, such as

> *. . . is married to . . . ,*

which relates two objects (people), or

> *. . . is the sum of . . . and . . . ,*

which relates three objects (numbers).

Gottlob Frege (1848–1925)

Predicate logic extends propositional logic, but the focus shifts from propositions to 'sentences' (the technical term is *formula*). This shift in focus is necessary because predicate logic allows you to form 'sentences' that are not necessarily true or false, and thus do not represent propositions. The construction rules comprise the propositional operations *and, or, not*, and the conditional (→), plus the two quantifiers, *all* and *some*. As in Aristotle's logic, 'some' is taken to mean 'at least one', as in the sentence "Some even number is prime." Another phrase that means the same is *There exists*, as in "There exists an even prime number."

The actual rules for sentence construction—the *grammar* of predicate logic—are a bit complicated to write down precisely and completely, but the following, simple examples should give the general idea.

In predicate logic, Aristotle's proposition *All men are mortal* is constructed like this:

For all x, if x is a man, then x is mortal.

This construction looks more complicated than the original version, and it is certainly more of a mouthful to say. The gain is that the proposition has been pulled apart into its constituents, exposing its internal, logical structure. This structure becomes more apparent when the logician's symbols are used instead of English words and phrases.

First, the logician writes the predicate *x is a man* in the abbreviated form *Man(x)* and the predicate *x is mortal* in the form *Mortal(x)*. This change of notation may, on occasion, make something simple appear complex and mysterious, but that is certainly not the intention. Rather, the aim is to direct concentration to the important patterns involved. The crucial aspect of a predicate is that it is true or false of one or more particular objects. What counts are (i) the property and (ii) the objects; everything else is irrelevant.

The proposition *Aristotle is a man* would thus be written *Man(Aristotle)*; the proposition *Aristotle is not*

a Roman would be written , ¬ *Roman(Aristotle)*; and the proposition *Susan is married to Bill* would be written *Married-to(Susan, Bill)*. This notation highlights the general pattern that a predicate is something that is true (or not) of certain objects:

$$\text{Predicate(object, object, . . .)}$$

$$\neg \text{ Predicate(object, object, . . .)}$$

Two further notations are used. The word *All*, or the phrase *For all*, is abbreviated by the symbol ∀, an upside-down letter A. The word *Some*, or the phrase *There exists*, is abbreviated by the symbol ∃, a back-to-front letter E. Using this notation, *All men are mortal* looks like this:

$$\forall x\colon Man(x) \rightarrow Mortal(x).$$

Written in this way, all the logical constituents of the proposition and the underlying logical pattern are immediately obvious:

(i) the quantifier, ∀;

(ii) the predicates, *Man* and *Mortal*;

(iii) the logical connection between the predicates, namely the conditional, →.

As a final example, the proposition *There is a man who is not asleep* looks like this:

$$\exists x\colon Man(x) \land \neg Asleep(x).$$

Though propositions written in this way look strange to anyone who sees such expressions for the first time, logicians have found the notation to be extremely valuable. Moreover, predicate logic is sufficiently powerful to express *all* mathematical propositions. It is certainly true that a definition or proposition expressed in predicate logic can appear daunting to the uninitiated. But this is because such an expression does not hide any of the logical structure; the complexity you see in the expression is the actual, structural complexity of the notion defined or of the proposition expressed.

As with the operations of propositional logic, there are rules that describe the 'algebraic' properties of the operations of predicate logic. For example, there is the rule

$$\neg\,[\forall x\colon P(x)] = [\exists x\colon \neg\,P(x)].$$

In this rule, $P(x)$ can be any predicate that is applicable to a single object, such as *Mortal(x)*. Written in English, an instance of this rule would be

[*Not all men like football*]

$= [$*Some men do not like football*$].$

Again, just as was the case with the rules for propositional logic, the equals sign has a special meaning; in this case, that the two expressions 'say the same thing'.

The development of predicate logic provided mathematicians with a means to capture, in a formal way, the patterns of mathematical proof. This is not to say that anyone ever advocated a slavish adherence to the rules of predicate logic. No one insists that all mathematical assertions be expressed in predicate logic itself, or that all proofs be formulated only in terms of modus ponens and the deduction rules that involve quantifiers (not stated here). To do so would, except for the simplest of proofs, be extremely laborious, and the resulting proof would be almost impossible to follow. But, by virtue of carrying out a detailed study of the patterns of predicate logic, mathematicians not only gained considerable understanding of the concept of a formal proof, they also confirmed that it was a reliable means of establishing mathematical truth. And *that* was of the highest importance in view of other developments taking place in mathematics at the same time, to which I turn next.

Abstraction and the Axiomatic Method

The second half of the nineteenth century was a glorious era of mathematical activity. In particular, it was during this period that mathematicians finally managed to work out a proper theory of the real-number continuum, thereby providing a rigorous foundation for the methods of the calculus that Newton and Leibniz had developed three hundred years earlier. (See Chapter 3.) Crucial to this progress was the increasing—and in many cases total—reliance on the axiomatic method.

All mathematics deals with abstraction. Though many parts of mathematics are motivated by, and can be used to describe, the physical world, the entities the mathematician actually deals with—the numbers, the geometric figures, the various patterns and structures—are pure abstractions. In the case of subjects such as the calculus, many of the abstractions involve the mathematical notion of 'infinity' and thus cannot possibly correspond directly to anything in the real world.

How does the mathematician decide if some assertion about these abstractions is true or not? The physicist, chemist, or biologist generally accepts or rejects a hypothesis on the basis of experiment, but most of the time the mathematician does not have this option. In cases that can be settled by straightforward, numerical computation, there is no problem. But in general, observations of events in the real world are at best suggestive of some mathematical fact, and on occasions can be downright misleading, with the mathematical truth being quite at variance with everyday experience and intuition.

The existence of nonrational real numbers falls into this category of counterintuitive mathematical facts. Between every two rational numbers lies a third, namely, the mean of the first two. On the basis of everyday experience, it would therefore seem reasonable to suppose that there is simply no room on the rational line for any more numbers. But as the Pythagoreans discovered to their immense dismay, this is not the case at all.

Although the Pythagoreans had been devasted by their discovery, mathematicians accepted the existence of irrational numbers because the existence had been *proved*. Ever since Thales, proof has been central to mathematics. In mathematics, truth is decided not by experiment, not by majority vote, and not by dictat—even if the dictator is the most highly regarded mathematician in the world. Mathematical truth is decided by proof.

This is not to say that proofs are all there is to mathematics. As the science of patterns, a lot of mathematical activity is concerned with finding new patterns in the world, analyzing those patterns, formulating axioms to describe them and facilitate further study, looking for the appearance in a new domain of patterns observed somewhere else, and applying mathematical theories and results to phenomena in the everyday world. In many of these activities, a reasonable question is, how well do the mathematical patterns and results accord with what may be physically observed, or with what may be computed? But as far as establishing mathematical truth is concerned, there is only one game in town: proof.

Mathematical truths are all fundamentally of the form

If A, then B.

This is because all mathematical facts are proved by deduction from some initial sets of assumptions, or *axioms* (from the Latin *axioma*, meaning 'a principle'). When a mathematician says that a certain fact (*B*) is 'true', what she means is that the 'fact' has been proved on the basis of some set of assumed axioms (*A*). It is permissible to express this result simply as "*B* is true" providing the axioms *A* are 'obvious', or at least widely accepted within the mathematical community. This point was the subject of an amusing commentary by the great English mathematician and philosopher Bertrand Russell, which is quoted in full in the box on the facing page.

For example, all mathematicians would agree that it is *true* that, between any positive whole number *N* and its double *2N*, there is a prime number. How can they be so sure? They certainly have not examined every possible case, since there are an infinite number of cases. Rather, this result has been *proved*. Moreover, the proof depends only on a set of axioms for the natural numbers that everyone accepts as definitive.

Provided you are sure that a proof is valid, the only part of this process that remains open to question is whether or not the axioms correspond to your intuitions. Once you write down a set of axioms, then anything you prove from those axioms will be mathematically true for the system of objects your axioms describe. (Strictly speaking, I should say "*any* system of objects," since most axiom systems will describe more than one system of objects, regardless of the purpose for which the axioms were originally formulated.) But it might well be that the system your axioms describe is not the one you set out to describe.

For instance, around 350 B.C., Euclid wrote down a set of axioms for the plane geometry of the world around us. From this set of axioms he was able to prove a great many results, results both aesthetically pleasing and immensely useful in everyday life. But, in the nineteenth century, it was discovered that the geometry described by Euclid's axioms might not be the geometry of the world around us. It might only be *approximately* right, albeit with a degree of approximation that is not noticeable in everyday life. In fact, present-day theories of physics assume geometries different from Euclid's. (This fascinating story is fully described in Chapter 4.)

By way of an illustration, here is a set of axioms formulated during the nineteenth century for the elementary arithmetic of the integers, the positive and negative whole numbers.

1. For all *m, n,*   *m + n = n + m* and *nm = mn*. (The commutative laws for addition and multiplication.)

Bertrand Russell and the Nature of Mathematical Truth

Writing about propositional logic in an article titled *Recent Work on the Principles of Mathematics*, published in the *International Monthly*, Vol 4 (1901), Bertrand Russell achieved both humor and complete accuracy in the following passage:

> Pure mathematics consists entirely of such asseverations as that, if such and such a proposition is true of *anything*, then such and such another

proposition is true of that thing. It is essential not to discuss whether the first proposition is really true, and not to mention what the anything is of which it is supposed to be true. . . . If our hypothesis is about *anything* and not about some one or more particular things, then our deductions constitute mathematics. Thus mathematics may be defined as the subject in which we never know what we are talking about, nor whether what we are saying is true.

2. For all m, n, k, $m + (n + k) = (m + n) + k$ and $m(nk) = (mn)k$. (The associative laws for addition and multiplication.)

3. For all m, n, k, $k(m + n) = (km) + (kn)$. (The distributive law.)

4. For all n, $n + 0 = n$. (The additive identity law.)

5. For all n, $1n = n$. (The multiplicative identity law.)

6. For all n, there is a number k, such that $n + k = 0$. (The additive inverse law.)

7. For any m, n, k, where $k \neq 0$, if $km = kn$, then $m = n$. (The cancellation law.)

These axioms are widely accepted by mathematicians as capturing important properties of the arithmetic of the integers. In particular, anything proved on the basis of these axioms will be described by a mathematician as 'true'. And yet, it is easy to write down 'facts' that no one has any hope of checking by direct computation or by experimental

processes such as counting piles of pennies. For instance, is the following identity true?

$$12{,}345^{678{,}910} + 314{,}159^{987{,}654{,}321} = 314{,}159^{987{,}654{,}321} + 12{,}345^{678{,}910}.$$

This identity is of the form $m + n = n + m$, so, on the basis of the axioms, you know it is 'true'. (In fact, the commutative law for addition *says* it is true; you don't even have to construct a proof in this case.) Is this a reliable way to 'know' something?

In writing down the above axioms for integer arithmetic, the mathematician is describing certain patterns that have been observed. Everyday experience with small numbers tells you that the order in which you add or multiply two numbers does not effect the answer. For example, $3 + 8 = 8 + 3$, as can be checked by counting out coins. If you count out three coins and then add a further eight, the number of coins you end up with is the same as if you first count out eight coins and then add a further three; in both cases you end up with eleven coins. This pattern is repeated with every pair of numbers you encounter. What is more, it seems reasonable to *assume* that the pattern will continue to be true for any other pair of numbers you might encounter, tomorrow, or the next day, or even for any pair of numbers anyone else might encounter at any time in the future. The mathematician takes this reasonable assumption, based on everyday experience, and *declares* it to be 'true', for *all* pairs of integers, positive or negative.

Because such rules are taken as axioms, any collection of objects that obeys all the rules listed above will have any property that can be proved on the basis of those axioms. For example, using the above axioms, it is possible to prove that the inverse of any 'number' is unique; that is, for any 'number' n, the 'number' k such that $n + k = 0$ is unique. Thus, whenever you have a system of 'numbers' that satisfies all these axioms, that system will never contain a 'number' that has two additive inverses.

The reason for the quotes around the word 'number' in the above paragraph is that, as mentioned earlier, whenever you write down a set of axioms, it generally turns out that there is more than one system of objects that satisfies those axioms. Systems that satisfy the above axioms are called integral domains.

Mathematicians have encountered a number of objects besides integers that form integral domains. For example, polynomial expressions form an integral domain. So too do certain of the finite arithmetics described in Chapter 1. In fact, the finite arithmetics modulo a prime number satisfy an additional axiom beyond those for an integral domain, namely:

8. For all n other than 0, there is a k such that $nk = 1$.

This is the multiplicative inverse law. It implies Axiom 7, the cancellation law. (More precisely, on the assumption of Axioms 1 through 6 and Axiom 8, it is possible to prove Axiom 7.) Systems that satisfy Axioms 1 through 8 are called fields.

There are many examples of fields in mathematics. The rational numbers, the real numbers, and the complex numbers are all fields. (See Chapter 3.) There are also important examples of fields where the objects are not 'numbers' in the sense most people would understand this word.

To the person who meets this kind of thing for the first time, it might all seem like a frivolous game—not that there is anything wrong with frivolous games! But the formulation of axioms and the deduction of various consequences of those axioms has, over the years, proved to be an extremely powerful approach to various kinds of phenomena, with many direct consequences for everyday life, both good and ill. Indeed, most of the components of modern life are based on the knowledge humankind has acquired with the aid of the axiomatic method. (That's not *all* they are based on, of course. But it is an essential component. Without the axiomatic method, technology, for instance, would have advanced little beyond that of a century ago.)

How is it that the axiomatic method has been so successful in this way? The answer is, in large part, because the axioms do indeed capture meaningful and correct patterns.

Which statements are accepted as axioms often depends as much on human judgment as anything else. For instance, on the basis of embarrassingly little concrete evidence, most citizens would be prepared to let their lives depend on the validity of the commutative law of addition. (How many times in your life have you bothered to check this law for a particular pair of numbers? Think of this the next time you step onto an airplane, where your life very definitely does depend on mathematics!)

There is certainly no logical basis for this act of faith. Mathematics abounds with examples of statements about numbers that are true for millions of cases, but not true in general. For example, the Mertens conjecture is a statement about natural numbers that had been verified by computer to be true for the first 7.8 billion natural numbers, before it was proved false in 1983. Yet, even before it had been proved false, no one had ever suggested adding this statement to the axioms for the natural numbers.

Why have mathematicians adopted the commutative law, for which the numerical evidence is slim, as an axiom, and left out other assertions for which there is a huge amount of numerical evidence? The decision is essentially a judgment call. For a mathematician to adopt a certain pattern as an axiom, not only should that pattern be a useful assumption, but it has to be 'believable', in keeping with her intuitions, and as simple as possible. Compared with these factors, supporting numerical evidence is of relatively minor importance—though a single piece of numerical evidence to the *contrary* would at once overturn the axiom!

Of course, there is nothing to prevent anyone from writing down some arbitrary list of postulates and proceeding to prove theorems from them. But the chances of those theorems having any practical application, or even any application elsewhere in mathematics, are slim indeed. Such an activity would be unlikely to find acceptance within the mathematical community. Mathematicians are not too disturbed if their work is described as 'playing games'. But they become really annoyed if it is described as a 'meaningless game'. And the history of civilization is very much on the mathematician's side: there is usually no shortage of applications.

The reason why the mathematician seeks the comfort of starting with a set of *believable* axioms for some system is that, once she starts trying to understand the system by proving consequences of those axioms, everything she does will rest upon those initial axioms. The axioms are like the foundations for a building. No matter how carefully the mathematician erects the walls and the rest of the structure, if the foundations are unsound, the entire construction may collapse. One false axiom, and everything that follows may be wrong or meaningless.

As outlined earlier, the initial step in the development of a new branch of mathematics is the identification of some pattern. Then comes the abstraction of that pattern to a mathematical object or structure, say the concept of a natural number or of a triangle. As a result of studying that abstract concept, the various patterns observed might lead to the formulation of axioms. At that point, there is

"Still Life with a Beer Mug," Fernand Léger, 1921. The increased abstraction in mathematics that took place during the early part of this century was paralleled by a similar trend in the arts. In both cases, the increased level of abstraction demands greater effort on the part of anyone who wants to understand the work.

no longer any need to know about the phenomenon that led to those axioms in the first place. Once the axioms are available, everything can proceed on the basis of logical proofs, carried out in a purely abstract setting.

Of course, the pattern that starts this whole process may be something that arises in the everyday world; for example, this is the case for the patterns studied in Euclidean geometry and, to some extent, in elementary number theory. But it is quite possible to take patterns that themselves arise from mathematics, and subject *those* patterns to the same process of abstraction. In this case, the result will be a new level of abstraction. The definition of integral domains is an example of this process of 'higher-level' abstraction. The axioms for an integral domain capture a pattern that is exhibited not only by the integers, but also by polynomials and a number of other mathematical systems, each of which is itself an abstraction, capturing other, 'lower-level' patterns.

During the nineteenth century, this process of abstracting from abstractions was taken to an extent that left all but a relatively small handful of professional mathematicians unable to appreciate most of the new developments in mathematics. Abstractions were piled on abstractions to form an enormous tower, a process that continues to this day. Though high levels of abstraction may cause many people to shy away from modern mathematics, increasing the level of abstraction does not in itself lead to more difficult mathematics. At each level of abstraction, the actual mechanics of doing mathematics remain much the same. Only the level of abstraction changes.

It is interesting to note that the trend toward increased abstraction over the past hundred years or so has not been unique to mathematics. The same process has been taking place in literature, in music, and in the visual arts. And often with a similar lack of appreciation by those not directly involved in the process.

Set Theory

As the level of abstraction in mathematics increased, mathematicians grew ever more dependent on the notion of an abstract *set*, 'set' being the name they adopted as a technical term to refer to any 'collection' of objects of some kind.

Georg Cantor (1845–1918).

New mathematical notions were being introduced and studied, such as groups, integral domains, fields, topological spaces, and vector spaces, and many of these were defined as certain sets of objects on which certain operations could be performed (operations such as 'addition' and 'multiplication' of various kinds).

Familiar old notions from geometry, such as lines, circles, triangles, planes, cubes, octahedra, and the like, were given new definitions as 'sets of points' satisfying various conditions.

And of course, Boole had developed his algebraic treatment of logic by regarding the syllogism in terms of sets.

The first complete, mathematical theory of abstract sets was worked out by the German mathematician Georg Cantor, toward the very end of the nineteenth century, though the beginnings of this theory were clearly present in the work of Boole. The basic ideas of Cantor's theory are presented in the box on the facing page.

Cantor's Set Theory

The basic ideas of Cantor's *set theory* can be found in Boole's treatment of the syllogism. The theory starts by developing an 'arithmetic' for sets.

If x and y are sets, let xy denote the set of all members common to x and y, and let $x+y$ denote the set consisting of all members of x together with all members of y.

The only difference between this definition and the one given earlier for Boole's logic is that the symbols x and y are now regarded as standing for *any* sets, not just sets arising from a logical proposition. The following 'arithmetical' axioms, given previously for Boole's classes, are true in this more general situation:

$$x + y = y + x \quad xy = yx$$
$$x + (y + z) = (x + y) + z \quad x(yz) = (xy)z$$
$$x(y + z) = xy + xz$$
$$x + 0 = x$$
$$x + x = x \quad xx = x$$

(Boole's axiom involving the object 1 does not appear, since there is no need for such an object in set theory, and introducing one can lead to technical problems.)

In present-day set theory, the set xy is called the intersection of x and y, and $x + y$ is called the union. A more common notation for these operations is $x \cap y$ for the intersection and $x \cup y$ for the union. In addition, contemporary mathematicians usually denote the empty set, the set having no members, by $\emptyset$ rather than 0. (The empty set is to set theory what the number zero is to arithmetic.)

For small sets, having no more than a dozen or so members, mathematicians use a notation that explicitly lists the members. For example, the set consisting of the numbers 1, 3, and 11 is written like this:

$$\{1, 3, 11\}.$$

Larger sets, or infinite ones, clearly cannot be depicted in this manner; in those cases, other ways have to be found to describe the set. If the members of a set have an obvious pattern, that pattern can be used to describe the set. For example,

$$\{2, 4, 6, \ldots\}$$

pretty obviously denotes the infinite set of all even natural numbers. Often, the only reasonable way is to describe the set in words, such as 'the set of all primes'.

Mathematicians write

$$x \in A$$

to denote that the object x is a member of the set A.

When mathematicians define some abstract mathematical object or system as a 'set of objects' satisfying certain properties, it usually does not matter what the members of the set are; rather, what counts are the operations that can be performed on those members. In fact, even that is not quite right. The real interest is in the *properties* of those operations. If they are studying the theory of the natural numbers, mathematicians do not need to know what a natural number really is—even assuming that question has a definitive answer. Nor do they need to know what addition and multiplication are. Their interest is in the various properties of numbers and of the operations of arithmetic, properties such as the commutativity of addition, whether one number divides evenly into another, whether a number is prime, and so forth. (Mind you, set theory does provide mathematicians with a way to answer those "What exactly are . . . ?" questions, as indicated in the box on pages 60–61.)

As a consequence of all of this effort, by the turn of the century set theory had become a general framework for an extremely large part of mathematics. So it was with considerable alarm that the world of mathematics woke up one June morning in 1902 to find that set theory was fundamentally inconsistent! That is to say, in Cantor's set theory it was possible to prove that 0 equals 1. (Strictly speaking, the problem arose in an axiomatization of set theory by Frege. But Frege's axioms simply formalized Cantor's ideas.)

Of all the things that can be wrong with an axiom system, inconsistency is absolutely the worst. It is possible to work with axioms that are hard to understand. It is possible to work with axioms that are counterintuitive. And all might not be lost if your axioms don't accurately describe the structure you intended to capture—maybe they will find some other application, as has happened on more than one occasion. But an inconsistent set of axioms is completely useless.

The inconsistency was found by Bertrand Russell, just as the second volume of Frege's treatise on his new theory was at the printer. Russell's argument was devastatingly simple.

According to Cantor and to Frege, and most likely to any mathematician at the time who cared to think about it, for any property, P, there will surely be a corresponding set of all those objects that have the property P. For example, if P is the property of being a triangle, then the set corresponding to P is the set of all triangles. (Frege's work amounted in large part to developing a formal theory of 'properties' to go with this idea—the predicate logic explored earlier in the chapter.)

Russell's argument concerns properties P that apply to sets. For such a property, the corresponding set is a set of sets. One property P that applies

to sets is that a set be a member of itself. Some sets have this property; they are members of themselves. For example, if M denotes the set of all sets that are explicitly named in this book, then M is a member of itself. ($M \in M$.) On the other hand, there are sets that do not have the property P; these sets are not members of themselves. For example, the set of all natural numbers is not itself a natural number, and hence is not a member of itself. ($\mathcal{N} \notin \mathcal{N}$.)

To arrive at a contradiction, Russell looked not at at the property P, but at the closely related property R that a set x is *not* a member of itself.

Some sets do have this property (for example, $\mathcal{N}$); others do not (M). Since R is a perfectly reasonable looking property, albeit a bit novel, according to Cantor and Frege there should be a corre-

Bertrand Russell (1872–1970).

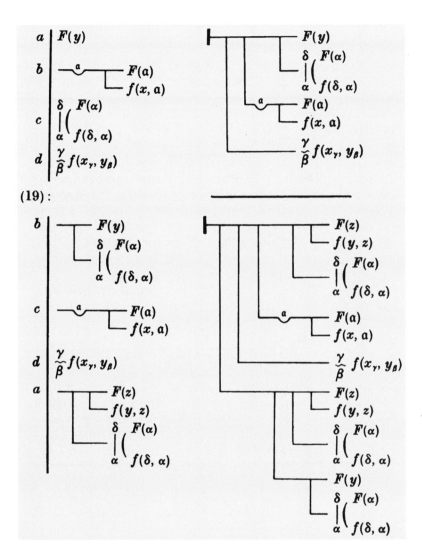

A page from Frege's treatise *Begriffsschrift, a formal language, modeled upon arithmetic, for pure thought*, published in 1879. This sample is typical of many of the eighty-two pages of the work.

sponding set, call it $\mathscr{C}$. $\mathscr{C}$ is the set of all those sets that have the property R. So far so good. All seems innocent enough, you might think.

Russell now asks the question, again perfectly reasonable, is this new set $\mathscr{C}$ a member of itself or is it not?

Russell points out that if $\mathscr{C}$ is a member of itself, then $\mathscr{C}$ must have the property R that defines $\mathscr{C}$. And what that means is that $\mathscr{C}$ is *not* a member of itself. So $\mathscr{C}$ both is and is not a member of itself, which is an impossible situation.

Russell continues by asking what happens if $\mathscr{C}$ is *not* a member of itself. In that case, $\mathscr{C}$ must fail to satisfy the property R. Therefore, it is not the case that $\mathscr{C}$ is not a member of itself. That last clause is a very complicated way of saying that $\mathscr{C}$ *is* a member of itself. So again, the inescapable conclusion is that $\mathscr{C}$ both is not and is a member of itself, an impossible state of affairs.

And now you are at a complete impasse. Either $\mathscr{C}$ is a member of itself or it is not. Either way, you finish by concluding that it both is and is not. This

Numbers from Nothing

With the growth in abstraction and the axiomatic method during the nineteenth century, the nature of the various objects in mathematics became far less important. Mathematical systems were typically defined as 'sets of objects', on which could be performed various operations, with the whole system satisfying various axioms. In many cases it was largely irrelevant exactly what those objects were.

This is not to say that mathematicians did not continue to ask themselves about the nature of the objects they studied, the numbers, lines, circles, planes, surfaces, geometric solids, and so forth. In fact, not only did they continue to ask such questions, for the first time in the history of mathematics they were able to provide definitive answers. And, ironically, those answers were presented in terms of the very thing that resulted in the answers being irrelevant to the study of the objects, namely, the notion of an abstract set!

A point in two-dimensional space may be defined as a certain *set of sets* of real numbers. The idea is to make use of Descartes' notion of coordinates, described in Chapter 4. A point in the plane has two coordinates, say the real numbers x and y. The 'point' is then defined to *be* the pair (x,y). The pair (x,y) is itself defined to be the *set*

$$\{\{x\},\{x,y\}\}.$$

This definition looks a little complicated, and it is. The idea is to define the notion of a pair or a 'point' so that if two points have the same first and second coordinates, then they are one and the same point, and vice versa. This definition does just that.

Having defined the notion of a point in the plane in terms of sets, a line or a plane figure may be defined as a certain *set* of points.

Real numbers can be defined as certain pairs of certain (infinite) sets of rational numbers. (This defines a real number as a *set of sets of sets* of rational numbers.)

Rational numbers may, in turn, be defined as certain (infinite) sets of pairs of integers. (So, a rational number is a *set of sets of sets* of integers.)

An integer may be defined as a pair of natural numbers, that is, as a *set of sets* of natural numbers.

And finally, how do you define a natural number? Does this tower of definitions go on forever, paralleling the physicist's seemingly neverending

result is known as Russell's paradox. Its discovery indicated that something was wrong with Cantor's set theory, but what?

Since Russell's reasoning was correct, there seemed to be only one way to resolve the paradox. The definition of the set $\mathscr{C}$ must be faulty in some way. And yet, it is just about as simple a definition of a set as anyone could hope for. The sets used to construct the various number systems are far more complicated. Unpalatable as it might be, therefore, it looked as though mathematicians had to jettison the assumption that for any property there will be a corresponding set—the set of objects having that property.

The situation was not unlike that facing the Pythagoreans when they discovered that there were lengths that did not correspond to any known numbers. And once again, there was no choice in the matter. Faced with a proof that something is fundamentally wrong, a theory has to be modified or replaced, regardless of how simple, elegant, or intuitive it is. Cantor's set theory had been all three: simple, elegant, and intuitive. But it had to be abandoned.

search for the ultimate building blocks of matter, or does it come to an end?

Provided you accept the notion of an arbitrary set, the unraveling process comes to an end. Moreover, it does so in a surprising way. In set theory, it is possible to construct the entire infinitude of natural numbers, starting with 'nothing', more precisely, with the empty set, $\emptyset$. The procedure goes like this.

Mathematicians define the number 0 *to be* the empty set.

The number 1 is then defined to be the set z, the set having exactly one member, that member being the number 0. (If you unravel that one step, you find that the number 1 is equal to the set $\{\emptyset\}$. If you think about it for a moment, you will realize that this is not quite the same as the empty set; $\emptyset$ has no members, whereas the set $\{\emptyset\}$ has one member.)

The number 2 is defined to be the set $\{0,1\}$. The number 3 is the set $\{0,1,2\}$. And so forth.

Each time a new number is defined, you use it, and all previous numbers, to define the next one. And the whole process starts out from the empty set, $\emptyset$, that is to say, from 'nothing'. Very clever.

What replaced Cantor's set theory was an *axiomatic* theory of sets developed by Ernst Zermelo and Abraham Fraenkel. Though Zermelo–Fraenkel set theory manages to remain close in spirit to Cantor's highly intuitive notion of an abstract 'set', and though it has proved to be an adequate foundation for all of pure mathematics, it has to be admitted that the theory is not particularly elegant. Compared to Cantor's theory, the seven axioms introduced by Zermelo and the subtle additional axiom introduced by Fraenkel form something of a motley collection of principles. They describe rules that give rise to

the various sets required in mathematics, while carefully skirting the kind of difficulty that Russell had uncovered.

Zermelo and Fraenkel's analysis of sets is sufficiently compelling that most mathematicians accept the resulting axioms as the 'right' ones on which to base mathematics. Yet for many, the first encounter with Russell's paradox and the steps required to circumvent it produces a sense of innocence lost. For all that the concept of a 'pure set' might seem the very essence of simplicity, closer analysis reveals otherwise. Set theory may well be an 'ultimate' pure creation of the human intellect, the essence of abstraction, but, as is the case with all great constructions of mathematics, it dictates its own properties.

Thirty years after Russell destroyed Cantor's intuitive set theory, a similar upheaval happened again, with equally devastating consequences. The victim on this second occasion was the belief in the axiomatic method itself, a belief that had found its most influential champion in the German mathematician David Hilbert.

Hilbert's Program and Gödel's Theorem

The axiomatic approach to mathematics made it possible for mathematicians to separate the issues of provability and truth. A mathematical proposition was *provable* if you could find a logically sound argument that deduced it from the appropriate axioms. A proposition that had been proved would be *true* provided the assumed axioms were true. The former notion, provability, was a purely technical one where the mathematician ruled supreme; the latter notion, truth, involved deep philosophical questions. By separating these two notions, mathematicians could sidestep thorny questions about the nature of truth and concentrate on the issue of proof. By restricting themselves to the task of proving results on the basis of an assumed axiom system, mathematicians were able to regard mathematics as a formal 'game', a game played according to the rules of logic, starting from the relevant axioms.

Discovering the appropriate axioms was clearly an important component of this 'formalistic' approach to mathematics, as it became known. Implicit in formalism was an assumption that, provided you looked long enough, you would eventually find all the axioms you needed in mathematics. In this way, the 'completeness' of the axiom system became a significant issue. Had enough axioms been found to be able to answer all questions? In the case of the natural numbers, for example, there was already an axiom system, formulated by Peano. Was this axiom system 'complete', or were additional axioms needed?

A second important question was, is a particular axiom system consistent? As Russell's paradox demonstrated all too clearly, writing down axioms that describe a highly abstract piece of mathematics is a difficult task.

The adoption of a purely formalistic approach to mathematics, involving as it did the search for consistent, complete axiom systems, became known as the Hilbert program, named after David Hilbert, one of the leading mathematicians of the time. Though he was not a 'logician' in the way that Frege and Russell were, questions about the foundations of mathematics were particularly important to Hilbert, whose own work was of a highly abstract nature. For instance, one of his legacies to mathematics is the 'Hilbert space', a sort of infinite-dimensional analogue of three-dimensional Euclidean space.

Any dreams that the Hilbert program could be completed were dashed in 1931, when a young Austrian mathematician called Kurt Gödel proved a result that was to change our view of mathematics forever. Gödel's theorem says that if you write down any consistent axiom system for some reasonably large part of mathematics, then that axiom system *must* be incomplete. There will always be some questions that cannot be answered on the basis of the axioms.

David Hilbert (1862–1943).

The phrase 'reasonably large' in the above paragraph is meant to exclude trivial examples such as 'the geometry of a single point' (0-dimensional geometry). In order to prove Gödel's theorem, you need to know that the axiom system concerned includes, or is rich enough to allow the deduction of, all the axioms of elementary arithmetic. This is probably as weak an assumption on axioms for mathematics as you will ever see.

The proof of Gödel's totally unexpected result is highly technical, but the idea is a simple one, having its origins in the ancient Greek paradox of the liar. The liar paradox refers to a person who stands up and says, "I am lying." If this assertion is true, then the person is indeed lying, which means that what he says is false. On the other hand, if the assertion is false, then the person must not be lying, so what he says is true. Either way the assertion is contradictory.

Gödel found a way to 'translate' this paradox into mathematics, replacing 'truth' by the notion of provability. His first move was to show how to translate propositional logic into number theory. Included in this translation process was the notion of a formal proof from the axioms. Then he produced a particular number-theoretic proposition that was analogous to the sentence

(*) The proposition on this page labeled with an asterisk is not provable.

First of all, proposition (*) must be either true or false (for the mathematical structure under consideration, say arithmetic of the natural numbers, or set theory). If the proposition is false, then it must be provable. You can see that by just looking at

Kurt Gödel (1906–1978).

what (*) says. But, since the consistency of the axioms is assumed, anything that is provable must be true. So, if it is false, then it is true, which is an impossible situation. Hence (*) must be true.

Is proposition (*) provable (from the axioms concerned)? Clearly not: its own truth implies that it is not provable, since that is what it says. Thus, proposition (*) cannot be proved from the given axioms.

So (*) is a proposition that is true for the structure but not provable from the axioms for that structure.

Gödel's argument can be carried through for any set of axioms you can write down for the structure. The stipulation that you must be able to 'write down' the axioms is important. After all, there is one trivial way to obtain an axiom system that can be used to prove all true propositions about the structure, and that is to declare the set of all true propositions *to be* the set of axioms. This is clearly not in the spirit of the axiomatic approach to mathematics, and is a quite useless axiomatization.

On the other hand, the phrase 'write down' can be taken in a very broad, idealistic sense. It allows not only for large finite sets of axioms, which can only be written down 'in principle', but for infinite sets of axioms of certain kinds. The key requirement is that you can stipulate one or more rules that show how the axioms *could* be written down. In other words, the axioms themselves must exhibit a very definite, linguistic pattern. Peano's axioms for the natural numbers and the Zermelo–Fraenkel axioms for set theory are both infinite axiom systems of this kind.

Gödel's result that, for important areas of mathematics such as number theory or set theory, no consistent set of axioms will be complete, clearly makes it impossible to achieve the goal of Hilbert's program. In fact, the situation is even 'worse'. Gödel went on to show that among the propositions that are true but unprovable from the axioms is one that asserts the consistency of those axioms. So there is not even the hope that you can prove that your axioms are consistent.

In short, in the axiomatization game, the best you can do is to *assume* the consistency of the axioms and *hope* that your axioms are rich enough to enable you to solve the problems of highest concern to you. You have to accept that you will be unable to solve all problems using your axioms; there will always be true propositions that you cannot prove from those axioms.

The Golden Age of Logic

Though it marked the end of the Hilbert program, the proof of Gödel's theorem ushered in what can only be described as the Golden Age of Logic. The period from around 1930 until the late 1980s saw intense activity in the area that became known by the general term 'mathematical logic'.

From the very beginning, mathematical logic split into several, connected strands.

Proof theory took to new lengths the study of mathematical proofs begun by Aristotle and continued by Boole. In recent years, methods and results from this branch of mathematical logic have found uses in computing, particularly in artificial intelligence.

Model theory, invented by the Polish-born American mathematician Alfred Tarski and others, investigated the connection between truth in a mathematical structure and propositions about that structure. The result alluded to earlier, that any axiom system will be true of more than one structure, is a theorem of model theory. In the 1950s, the American logician and applied mathematician Abraham Robinson used techniques of model theory to work out a rigorous theory of 'infinitesimals', thereby providing a way to develop the calculus (see Chapter 3) quite different from the one worked out in the nineteenth century.

Set theory took on new impetus when model-theoretic techniques were brought to bear on the study of Zermelo–Fraenkel set theory. A major breakthrough came in 1963, when Paul Cohen, a young American mathematician, found a means to prove, rigorously, that certain mathematical statements were *undecidable*, that is to say, could be proved neither true nor false from the Zermelo–Fraenkel axioms. This result was far more wide-ranging than Gödel's theorem. Gödel's theorem simply tells you that, for an axiom system such as that of Zermelo–Fraenkel set theory, there will be *some* undecidable statements. Cohen's techniques enabled mathematicians to take certain, specific mathematical statements and prove that those particular statements were undecidable. Cohen himself used the new technique to answer the continuum problem, a famous question posed by Hilbert in 1900. Cohen showed that the question was not decidable.

Computability theory also began at around the time of Gödel, and, indeed, Gödel himself made major contributions to this field. From today's viewpoint, it is interesting to look back at the work on the concept of computability carried out in the 1930s, two decades before any kind of 'real' computer would be built, and fifty years before the arrival of today's desktop computers. In particular, Alan Turing, an English mathematician, proved an abstract theorem

Paul Cohen of Stanford University.

that established the theoretical possibility of a single computing machine that could be 'programmed' to perform any computation you like. The American logician Stephen Cole Kleene proved another abstract theorem showing that the program for such a machine was essentially no different from the data it would run on.

All these areas of mathematical logic had in common the characteristic that they *were* mathematical. It was not just that the studies were carried out in a mathematical fashion, but that, by and large, their subject matter was mathematics itself. Thus, the enormous advances made in logic during this century were achieved at a price. Logic had begun with Aristotle's attempts to analyze human reasoning in general, not just reasoning about mathematics. Boole's algebraic theory of logic brought the methods of mathematics to the study of reasoning, but the patterns of reasoning examined were arguably still general ones. However, the highly technical mathematical logic of the twentieth century was exclusively mathematical, both in the techniques used and the kinds of reasoning studied. In achieving mathematical perfection, logic had, to a large extent, broken away from its original goal of using mathematics to describe patterns of the human mind.

But while logicians were developing mathematical logic as a new branch of mathematics, the use of mathematics to describe patterns of the mind was being taken up once again. This time it was not the mathematicians who were doing the work, but a quite different group of scholars.

Patterns of Language

To most people, it comes as a shock to discover that mathematics may be used to study the structure of language—the real, human languages they use in their everyday lives, English, Spanish, Japanese, and so on. Surely, ordinary language is not in the least mathematical, is it?

Take a look at A, B, and C below. In each case, without hesitation, decide whether you think that what you see is a genuine sentence of English.

A. Biologists find the A-spinelli morphenium an interesting species to study.

B. Many mathematicians are fascinated by quadratic reciprocity.

C. Bananas automobile because mathematics specify.

Almost certainly you decided, without having to give the matter any thought at all, that A and B are proper sentences but that C is not.

And yet A involves some words that you have never seen before. How can I be so sure? Because I made up the two words 'spinelli' and 'morphenium'. So in fact, you happily classified as a sentence of English a sequence of 'words', some of which are not really words at all!

In the case of example B, all the words are indeed genuine English words, and the sentence is in fact true. But unless you are a professional mathematician, you are unlikely to have ever come across the phrase 'quadratic reciprocity'. And yet again, you are quite happy to declare B to be a genuine sentence.

On the other hand, I am sure you had no hesitation deciding that C is not a sentence, even though in this case you were familiar with all the words.

How did you perform this seemingly miraculous feat with so little effort? More precisely, just what is it that distinguishes examples A and B from example C?

It obviously has nothing to do with whether the sentences are true or not, or even with whether you understand what they are saying. Nor does it make any difference whether you know all the words in the sentence, or even if they are genuine words or not. What counts is the overall *structure* of the sentence (or nonsentence, as the case may be). That is to say, the crucial feature is the way the words (or nonwords) are put together.

A Grammar for a Fragment of English

Using mathematical notation suggested by Noam Chomsky, the rules of grammar for the English language are written in the following fashion:

$$DNP\ VP \to S$$

$$V\ DNP \to VP$$

$$P\ DNP \to PP$$

$$DET\ NP \to DNP$$

$$DNP\ PP \to DNP$$

$$A\ NP \to NP$$

$$N \to NP$$

In words, the first of these rules says that a definite noun phrase (*DNP*) followed by a verb phrase (*VP*) gives you a sentence (*S*); the second says that a verb (*V*) followed by a *DNP* gives you a *VP*; the third that a preposition (*P*) followed by a *DNP* gives you a prepositional phrase (*PP*); the next that a determiner (*DET*), such as the word 'the', followed by a noun phrase (*NP*) gives you a *DNP*.

Given that A stands for adjective and *N* stands for noun, you can figure out the meaning of the last three rules for yourself.

In order to use the grammar to generate (or analyze) sentences of English, all you need is a lexicon, a list of words, together with their linguistic categories. For example:

$$\text{to} \to P$$

$$\text{runs} \to V$$

$$\text{big} \to A$$

$$\text{woman} \to N$$

$$\text{car} \to N$$

$$\text{the} \to DET$$

Using this grammar, it is possible to analyze the structure of the English sentence

The woman runs to the big car.

Such an analysis is most commonly represented in the form of a *parse tree*, as shown in the figure on

This structure is, of course, a highly abstract thing; you can't point to it the way you can point to the individual words or to the sentence. The best you can do is observe that examples A and B *have* the appropriate structure, but example C *does not*. And that is where mathematics comes in, for mathematics is the science of abstract structure.

The abstract structure of the English language that we rely upon, subconsciously and effortlessly, in order to speak and write to each other and to understand each other, is what is called the syntactic structure of English. A set of 'axioms' that describes that structure is called a grammar for the language.

This way of looking at languages is relatively recent, and was inspired by the work in mathematical logic during the 1930s and 1940s.

Around the turn of the century, there was a shift from studies of the historical aspects of languages, their roots and evolution (often referred to as historical linguistics or philology), to an analysis of languages as *communicative systems* as they exist at any given point of time, regardless of their evolution. This kind of study is generally referred to as synchronic linguistics. Modern, mathematically based linguistics grew out of this development. The change from historical linguistics to the study of this page. (In so far as it resembles a real tree, a parse tree is upside down, with its 'root' at the top.)

At the 'top' of the tree is the sentence. Then, each move you make from any point in the tree, down by one level, indicates the application of one rule of the grammar. For example, the very first step down, starting from the topmost point, represents an application of the grammar rule

$$DNP\ VP{\rightarrow}S.$$

The parse tree represents the abstract structure of the sentence. Any competent English speaker is able to recognize (generally subconsciously) such a structure. You may replace each of the words in this tree with other words, or even nonwords, and, provided your substitutions 'sound right' for each grammatical category, the resulting sequence of words will sound like an English sentence. By providing axioms that determine all such parse trees, the formal grammar thus captures some of the abstract structures of English sentences.

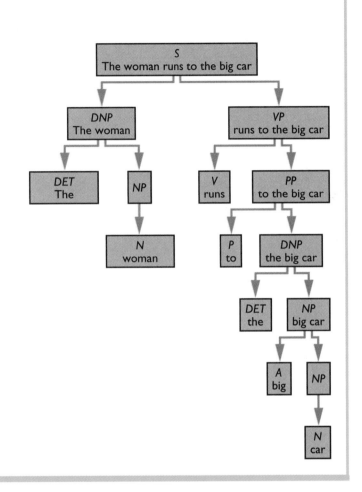

language as a *system* was due essentially to Mongin-Ferdinand de Saussure in Europe and to Frank Boas and Leonard Bloomfield in the United States.

Bloomfield, in particular, emphasized a scientific approach to linguistics. He was an active proponent of the philosophical position known as 'logical positivism', advocated by the philosopher Rudolf Carnap and the Vienna Circle. Inspired by the recent work in logic and the foundations of mathematics, in particular the Hilbert program, logical positivism attempted to reduce all meaningful statements to a combination of propositional logic and sense data (what you can see, hear, feel, or smell). Some lin-

guists, in particular the American Zellig Harris, went even further than Bloomfield, suggesting that *mathematical* methods could be applied to the study of language.

The process of finding 'axioms' that describe the syntactic structure of language was begun by the American linguist Noam Chomsky, though the idea for such an approach had been proposed over a century earlier, by Wilhelm von Humboldt. "To write a grammar for a language," Chomsky suggested, "is to formulate a set of generalizations, i.e., a theory, to account for one's observations of the language."

Chomsky's revolutionary new way to study language was described in his book *Syntactic Structures*, published in 1957. Within a couple of years of its appearance, this short treatise—the text itself occupies a mere 102 pages—transformed American linguistics, turning it from a branch of anthropology into a mathematical science. (The effect in Europe was less dramatic.)

The box on the previous page gives a small fragment of a Chomsky-style grammar for the English language. Of course, English is very complex, and this example gives just a few of the rules of English grammar. But this is surely enough to indicate the mathematical nature of the structure captured by a grammar.

Having found a way to use mathematical techniques in the study of language, Chomsky took his analysis a step further, examining the nature of formal grammars in general. By placing various restrictions on the kinds of rules that may appear in a grammar, he introduced a whole hierarchy of grammars, known as the Chomsky hierarchy. The grammars in this hierarchy ranged from the simplest, called regular grammars, to the most complex, the phrase-structure grammars.

The regular grammars describe only very simple 'languages'. One example is the numerical language you use to release a combination lock. You probably have never thought of a combination lock as having anything to do with language, but it does. The 'sentences' in the language are sequences of numbers. 'Grammatical sentences' are sequences that trigger the lock to open. For most combination locks, there is only one grammatical sentence.

At the other end of Chomsky's spectrum, the phrase-structure grammars are fairly complicated, and describe large parts of human-language syntax.

Though mathematical techniques are mostly used descriptively in linguistics, rather than to 'prove theorems about language', it is possible to prove theorems about the mathematics itself, and that also provides insight into the nature of lan-

Who Wrote the Federalist Papers?

Who wrote the various *Federalist* papers? Students of the origins of the United States Constitution have an interest in this question. Mathematics provided the answer.

The *Federalist* is a collection of eighty-five papers published during the period from 1787 to 1788, by Alexander Hamilton, John Jay, and James Madison. Their goal was to persuade the people of New York State to ratify the new Constitution. Because none of the individual papers bore the name of its actual author, the Constitutional historian was faced with a problem: just who did write each paper? The question has considerable interest, since the papers provide insights into the men who formulated the Constitution, and framed the future of the United States. An answer was not obtained until 1962, when the American mathematicians Frederick Mosteller and David Wallace used mathematics to resolve the issue.

For all but 12 of the papers, historical evidence had already provided the answer. It was generally agreed that 51 had been written by Hamilton, 14 by Madison, and 5 by Jay. That left 15 unaccounted for. Of these, 12 were in dispute between Hamilton and Madison and 3 were believed to be joint works.

Mosteller and Wallace's strategy was to look for patterns in the writing—not the syntactic patterns studied by Chomsky and other linguists, but numerical patterns. What made this approach possible is the fact that each individual has a distinctive style of writing, elements of which are susceptible to a statistical analysis. To determine authorship, various numerical values associated with the disputed paper could be compared with other writings known to have come from the pens of the persons concerned.

One obvious number to examine is the average number of words an author uses in a sentence.

Though this number can vary depending on the topic, when an author writes on a single topic, as is the case with the *Federalist* papers, the average sentence length remains remarkably constant from paper to paper.

In the *Federalist* case, however, this approach was far too crude. For the undisputed papers, Hamilton averaged 34.5 words per sentence and Madison averaged 34.6. It was impossible to tell the two authors apart simply by looking at sentence length.

A number of seemingly more subtle approaches, such as comparing the use of 'while' against 'whilst', also failed to lead to a definite conclusion.

What finally worked was to compare the relative frequency with which each author used each of thirty carefully chosen common words, 'by', 'to', 'this', 'there', 'enough', 'according', and so forth. When the rates at which the three authors used each of these words were subjected to a computer analysis that looked for numerical patterns, the results were quite dramatic. Each author's writing exhibited a distinctive numerical 'fingerprint'.

For example, in his undisputed writings, Hamilton used 'on' and 'upon' almost equally, at the rate of about 3 times per 1,000 words. In contrast, Madison almost never used 'upon'. Hamilton used the word 'the' on average 91 times per 1,000 words and Madison 94 times per 1,000 words, which does not distinguish between them, but Jay's rate was 67 per 1,000, so the frequency of use of the word 'the' can help to distinguish between Jay and the other two. The chart shows the distribution of rates of occurrence of the word 'by' in a varied collection of 48 papers by Hamilton, 50 papers by Madison, and the 12 disputed *Federalist* papers.

Taken on its own, the evidence of any one such word is suggestive, but hardly convincing. However, the detailed statistical analysis of the word rates carried out for all thirty words was far more reliable, and the possibility of error in the final conclusion was provably miniscule.

The conclusion was that, almost certainly, Madison had authored the disputed papers.

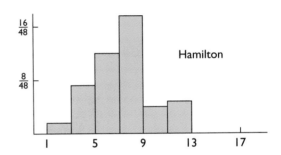

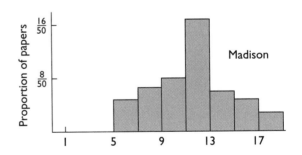

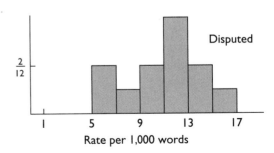

Noam Chomsky of the Massachusetts Institute of Technology.

('Understood' in this context means that the device will make an appropriate response to an input that is grammatical according to the particular grammar concerned. For example, in the case of a combination lock, the appropriate response is that the lock is released.)

Next in complexity are computers that have a fairly rudimentary memory (a single 'stack', for those who know what this means). The languages these computers 'understand' are precisely those determined by the context-free grammars, a class of grammars in Chomsky's hierarchy of particular interest to linguists.

Add still more memory to your computer and you find that the languages 'understood' are exactly those determined by a context-sensitive grammar. The computers in this case are known as linear-bounded automata.

Add more memory still to your computing device, and you obtain what is known as a Turing machine. This hypothetical computer was introduced by the logician Alan Turing in 1935, in an attempt to capture the patterns of thought involved in human *computation* (rather than the human deduction examined by previous generations of logicians). The Turing machine is a device having an unlimited, though rudimentary, memory. During the period from the 1930s to the 1950s, a large number of results demonstrated that, for all its simplicity and hypothetical nature, a Turing machine could, in principle, and given enough time, perform any computation that could ever be performed by any kind of computer, no matter how complex.

Chomsky tied language to computation in a very strong way when he proved that the languages that can be 'understood' by a Turing machine are just those whose grammar is a so-called phrase-structure grammar, a particularly important class of grammars in linguistics.

Of course, the success of Chomsky's work does not mean that mathematics is able to capture everything there is to know about language. Mathematics never captures all there is to know about anything. Understanding gained through mathematics is just a part of a much larger whole. A human lan-

guage. For instance, Chomsky proved a number of theorems that indicate fundamental connections between the grammars in his hierarchy and various kinds of hypothetical computing device studied by logicians and computer scientists.

The simplest of these hypothetical computers is the so-called finite automaton. Roughly speaking, this is a computer that can respond to input but has no memory. Chomsky showed that the languages that could be 'understood' by such a device are precisely those whose grammar is a regular grammar.

guage such as English is a highly complex system, constantly changing and evolving. A grammar captures only one part of a much larger picture, but it is an important part. It is also one of the parts (there are others) that are best handled using mathematical techniques. English syntax is a complex, abstract structure, and mathematics is simply the most precise intellectual tool there is to describe abstract structures.

Many of the abstract structures captured by mathematics arise in the physical world. Some arise within mathematics itself—structures built from abstract mathematical objects, the 'patterns of patterns'. But there are other kinds of abstract structure, and among them are the structures of everyday language. When a linguist writes down some of the rules of grammar, what she is doing is describing, by means of symbols on a page, the abstract structure of the English language that lives in our minds.

Marcel Duchamp, *Nude Descending a Staircase #2* **(1912)**

Motion and Change

We live in a world that is in constant motion, much of it recognizably regular.

The sun rises each morning and sweeps a steady path across the daytime sky. As the seasons pass, the height of the sun's path above the horizon rises and falls, again in a regular manner.

On the surface of the earth, a dislodged rock will roll down the hillside, and a rock thrown in the air will curve through the air before falling to the ground.

Moving air brushes against our faces, rain falls on our heads, the tides come in and go out, the clear sky fills with drifting clouds, animals run and walk or fly or swim, plants spring from the ground, grow, and then die, diseases break out and spread through populations.

Motion is everywhere, and without it there would be no such thing as life. 'Still life' exists only in the art gallery; it is not real life, for motion and change of one kind or another are the very essence of life.

Some motions appear chaotic, but much has order and regularity, exhibiting the kind of regular pattern that is, or at least ought to be, amenable to mathematical study. But the tools of mathematics are essentially *static*; numbers, points, lines, equations, and so forth do not in any way incorporate motion. So, in order to study motion, a way has to be found whereby these static tools can be brought to bear on patterns of change. It took some two thousand years of effort for humankind to achieve this feat, and the biggest single step was the development of the calculus in the middle of the seventeenth century. This one mathematical advance marked a turning point in human history, having as dramatic and revolutionary an effect on our lives as the invention of the wheel or of the printing press.

In essence, the calculus consists of a collection of methods to describe and handle patterns of infinity—the infinitely large and the infinitely small. For, as the ancient Greek philosopher Zeno indicated by a series of tantalizing paradoxes (see presently), the key to understanding the nature of motion and change is to find a way to tame infinity.

There is another paradox here. Though infinity is not part of the world we live in, it seems that the human mind requires a mastery of infinity in order to analyze motion and change in the world. Perhaps then the methods of the calculus say as much about ourselves as they do about the physical world to which they can be applied with such effect. The patterns of motion and change captured using the calculus certainly correspond to the motion and change we observe in the world, but, as patterns of infinity, their existence is inside our minds. They are patterns we human beings develop to help us comprehend the world.

The Paradox of Motion

For the most part, mathematicians have concentrated on continuous, rather than discrete, motion. But on first analysis, the very idea of continuous motion seems to be paradoxical. For consider: at a particular instant of time, any object must be at a particular location, a certain position in space. At that instant, the object is indistinguishable from a similar object at rest. But this will be true of *any* instant of time, so how can the object move? Surely, if the object is at rest at every instant, then it is always at rest.

This particular paradox of motion was first put forward by the Greek philosopher Zeno, probably as an argument against the numerical-based mathematical studies of the Pythagoreans. Zeno, who lived about 450 B.C., was a student of Parmenides, the founder of the Eleatic school of philosophy that flourished for a while in Elea, in Magna Graecia. Expressed originally in terms of an arrow in flight, Zeno's puzzle is a genuine paradox if one regards space as consisting of a multiplicity of adjacent 'points' and time as a succession of 'instants'.

The arrow paradox strikes a blow against the view of space and time as atomic. Another of Zeno's puzzles creates a paradox for those who believe that space and time are not atomic but infinitely divisible. This is the paradox of Achilles and the tortoise,

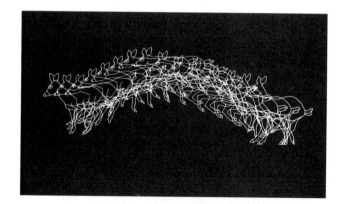

The paradox of motion. At any instant an object must be at rest, an idea captured by this illustration of a leaping deer. Since this is true for all instants, the object will always be at rest, so how can motion arise? The Greek philosopher Zeno posed this paradox as a challenge to the belief that time consisted of a succession of discrete instants.

Number Patterns in Motion

A simple experiment illustrates a particularly striking number pattern that arises in motion. Take a long length of plastic guttering and fix it to form a descending ramp, as shown in the illustration. Place a ball at the top end and release it. Mark off the position of the ball after it has rolled for exactly one second (say). Now mark off the entire length of guttering into lengths equal to the first, and number the markings 1, 2, 3, and so on. If you now release the ball from the top again, and follow its descent, you will notice that after 1 second it has reached mark number 1, after 2 seconds it is at

mark number 4, after 3 seconds it is at mark number 9, and, if your ramp is long enough, after a further second's descent it is at mark 16.

The pattern here is obvious: after n seconds' descent, the ball is exactly at the mark numbered n^2. And what is more, this is true *regardless of the angle at which you incline the ramp.*

Though simple to observe, a complete mathematical decription of this pattern requires the full power of the differential and integral calculus, techniques to be explained in the remainder of this chapter.

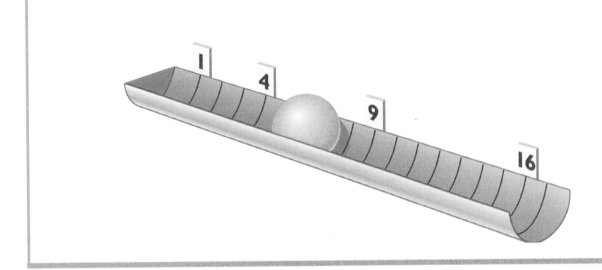

perhaps the best known of Zeno's arguments. Achilles is to race the tortoise over, say, 100 meters. Since Achilles can run ten times faster than the tortoise, the tortoise is given a 10-meter start. The race starts and Achilles sets off in pursuit of the tortoise. In the time it takes Achilles to cover the 10 meters needed to reach the point from where the tortoise had started, the tortoise has covered exactly 1 meter, and so is 1 meter ahead. By the time Achilles has covered that extra meter, the tortoise is a tenth of a meter in the lead. When Achilles gets to that

point, the tortoise is a hundredth of a meter ahead. And so on, ad infinitum. Thus, the argument goes, the tortoise remains forever in the lead, albeit by smaller and smaller margins; Achilles never overtakes his opponent to win the race.

The purpose of these paradoxes was certainly not to argue that an arrow cannot move or that Achilles can never overtake the tortoise. Both of these are undeniable, empirical facts. Rather Zeno's puzzles presented challenges to the attempts of the day to provide analytic explanations of space, time, and motion, challenges that the Greeks themselves were not able to meet. Indeed, truly satisfactory resolutions to the paradoxes were not found until the end of the nineteenth century, when mathematicians finally came to grips with the mathematical infinite.

Infinite Series

The key to the eventual development of a mathematical treatment of motion and change was to find a way to handle infinity. And that meant finding ways to describe and manipulate the various *patterns* that involve infinity.

For example, Zeno's paradox of Achilles and the tortoise may be disposed of once you have a way of dealing with the pattern involved. The amounts by which the tortoise is ahead of Achilles at each stage of the analysis given in the problem are (in meters) $10, 1, \frac{1}{10}, \frac{1}{100}, \frac{1}{1000}$, and so on. Thus the 'paradox' hinges on what we make of the infinite 'sum'

$$10 + 1 + \frac{1}{10} + \frac{1}{100} + \frac{1}{1000} + \cdots,$$

where those three dots signify that this 'sum' goes on forever, *following the pattern indicated*.

There is no hope of actually adding together all the infinitely many terms in this 'sum'. Indeed, I can't even write it out in full, which is why I keep enclosing the word 'sum' in quotes; it is not a sum in the normal sense of the word. In fact, to avoid this continual use of quotes, mathematicians refer to such infinite 'sums' as infinite *series*. This is one of a number of occasions when mathematicians take an everyday word and give it a technical meaning, often only slightly related to its everyday use.

By shifting attention from the individual terms in the series to the overall pattern, it is easy to find the value of the series. Let S denote the unknown value:

$$S = 10 + 1 + \frac{1}{10} + \frac{1}{100} + \frac{1}{1000} + \cdots.$$

The pattern in this series is that each successive term is one-tenth the previous term. So, if you multiply the entire series through by $\frac{1}{10}$, you obtain the same series again, apart from the first term:

$$\frac{1}{10}S = 1 + \frac{1}{10} + \frac{1}{100} + \frac{1}{1000} + \cdots.$$

If you now subtract this second identity from the original one, all the terms on the right-hand side cancel out in pairs, apart from the initial 10 in the first series:

$$S - \frac{1}{10}S = 10.$$

Now you have a *finite* equation, which can be solved in the usual way:

$$\frac{9}{10}S = 10,$$

so

$$S = \frac{100}{9}.$$

In other words, Achilles draws level with the tortoise after he has covered *exactly* $11\frac{1}{9}$ meters.

The crucial point is that an infinite series can have a finite value; Zeno's puzzle is only paradoxical if you think that an infinite series must have an infinite value.

Notice that the key to finding the value of the series was to shift attention from the process of adding the individual terms to the identification and manipulation of the overall *pattern*. In a nutshell, this is the key to handling the infinite in mathematics.

Of course, it is not entirely clear that one may justifiably 'multiply through' an infinite series by a fixed number, as I did above, or that one may then subtract one series from another, term by term, as I also did. Infinite patterns are notoriously slippery customers, and it is easy to go wrong. Take a look at the following infinite series, for instance:

$$S = 1 - 1 + 1 - 1 + 1 - 1 + \cdots .$$

If you multiply through by -1, you obtain the same series 'shifted one along':

$$S = 1 - 1 + 1 - 1 + 1 - 1 + \cdots$$
$$-S = \quad - 1 + 1 - 1 + 1 - 1 + \cdots .$$

If you then subtract the second series from the first, all terms on the right cancel out, apart from the first term of the top series, to leave

$$2S = 1.$$

The conclusion is that $S = \frac{1}{2}$.

All well and good, you might think. But now suppose that you take the original series and pair off the terms like so:

$$S = (1 - 1) + (1 - 1) + (1 - 1) + \cdots .$$

Again, this seems a perfectly reasonable thing to do to the overall pattern; though there are infinitely many terms in the series, I have described the *pattern* whereby the bracketing is done. But this time, each bracketed pair works out to be 0, so now the conclusion is that

$$S = 0 + 0 + 0 + \cdots ,$$

which means that $S = 0$.

Or you could apply brackets according to the following pattern:

$$S = 1 + (-1 + 1) + (-1 + 1) + (-1 + 1) + \cdots .$$

This time you obtain the value $S = 1$.

The original series for S was given by a perfectly understandable pattern. I showed how to manipu-late it in three different ways, using three different patterns of manipulation, and arrived at three different answers: $S = \frac{1}{2}$, 0, 1. Which answer is correct?

In fact, there is no correct answer. The pattern in the second example cannot be handled in a math-ematical fashion: this particular infinite series sim-ply does not have a 'value'. On the other hand, the series that arises from Achilles and the tortoise does have a value, and indeed, the pattern manipulation I carried out in that case is permissible. Sorting out the distinction between series that can be manipu-lated and those that cannot, and developing a sound theory of how to handle infinite series, took hun-dreds of years of effort, and was not completed un-til late in the nineteenth century.

A particularly elegant illustration of the man-ner in which the value of an infinite series may be determined by manipulating the pattern of the se-ries is provided by the so-called geometric series. These are series of the form

$$S = a + ar + ar^2 + ar^3 + \cdots ,$$

where each successive term is obtained by multi-plying the previous term by some fixed amount r. Geometric series arise frequently in everyday life, for example in radioactive decay and in the compu-tation of the interest you must pay on a bank loan or a mortgage. As it happens, the series that arose in the paradox of Achilles and the tortoise was also such a series. In fact, the method I used to find the value of that series works for any geometric series. To obtain the value of S, you multiply the series through by the common ratio r and subtract the new series from the first. All terms cancel out in pairs apart from the initial term a in the first series, leav-ing the equation

$$S - rS = a.$$

Solving this equation for S you obtain

$$S = \frac{a}{1 - r}.$$

Off to infinity. The resolution of Zeno's paradoxes required the development of a theory of infinite processes. The theory of such processes also led to the discovery of the calculus by Newton and Leibniz.

So, for example, the series

$$1 + \frac{1}{2} + \frac{1}{4} + \frac{1}{8} + \cdots + \frac{1}{2^n} + \cdots$$

has initial term $a = 1$ and ratio $r = \frac{1}{2}$, so its value is

$$\frac{1}{1 - \frac{1}{2}} = 2.$$

Obviously, one consequence of the ratio r being less than 1 (and more than -1 in the case where r is negative) is that the terms in the series get smaller as you go out. Could this be the crucial factor that enables you to find a finite value to an infinite series?

On the face of it, this hypothesis seems reasonable; if the terms get progressively smaller, their effect on the sum becomes increasingly insignificant. If this is indeed the case, then it will follow that the following elegant series has a finite value:

$$1 + \frac{1}{2} + \frac{1}{3} + \frac{1}{4} + \frac{1}{5} + \cdots .$$

Because of its connection to certain patterns on the musical scale, this series is known as the harmonic series.

If you add together the first thousand terms, you obtain the value 7.485 (to three places of decimals); the first million terms add together to give 14.357 (to three places); the first billion give approximately 21, and the first trillion about 28. But what is the value of the entire, infinite sum?

The answer is that there is no value, a result first discovered by Nicolae Oresme in the fourteenth century. Thus, the fact that the terms of an infinite series get progressively smaller is not in itself enough to guarantee that the series has a finite value.

How do you set about proving that the harmonic series does not have a finite value? Certainly not by adding together more and more terms. Sup-

The only remaining question is whether or not the various manipulations just described are valid or not. A more detailed examination of the pattern indicates that the manipulations are permissible in the case where r is less than 1 (in the case of a negative r, it must be greater than -1), but are not valid for other values of r.

pose you were to start to write out the series term by term on a ribbon, allowing one centimeter for each term (a gross underestimation, since you would need to write down more and more digits the further along the series you went). You would need some 10^{43} centimeters of ribbon to write down enough terms to sum to a value that exceeds 100. But 10^{43} centimeters is about 10^{25} light-years, which exceeds the known size of the entire universe (for which 10^{12} light-years is one current estimate).

The way to show that the harmonic series has an infinite value is to work with the pattern, of course. Start off by observing that the third and fourth terms are both at least $\frac{1}{4}$, so their sum is at least $2 \times \frac{1}{4} = \frac{1}{2}$. Now notice that the next four terms, namely $\frac{1}{5}, \frac{1}{6}, \frac{1}{7}, \frac{1}{8}$, are all at least $\frac{1}{8}$, so their sum is at least $4 \times \frac{1}{8} = \frac{1}{2}$. Likewise, the next sixteen terms, from $\frac{1}{9}$ to $\frac{1}{32}$, are all at least $\frac{1}{32}$, so they also add up to at least $16 \times \frac{1}{32} = \frac{1}{2}$.

By taking increasingly longer groups of terms, according to the pattern 2 terms, 4 terms, 8 terms, 16 terms, 32 terms, and so on, you keep on getting sums that in each case are at least $\frac{1}{2}$. This procedure will lead to infinitely many repetitions of $\frac{1}{2}$, and adding together infinitely many $\frac{1}{2}$s will surely produce an infinite result. But the value of the harmonic series, if there is indeed a value, will be at least as big as this infinite sum of $\frac{1}{2}$s. Hence the harmonic series cannot have a finite value.

During the seventeenth and eighteenth centuries, mathematicians became ever more skilled at manipulating infinite series. For instance, the Scotsman James Gregory found the following result in 1671:

$$\frac{\pi}{4} = \frac{1}{1} - \frac{1}{3} + \frac{1}{5} - \frac{1}{7} + \cdots .$$

In 1736, Euler discovered another infinite series whose value involved π:

$$\frac{\pi^2}{6} = \frac{1}{1^2} + \frac{1}{2^2} + \frac{1}{3^2} + \frac{1}{4^2} + \cdots .$$

In fact, Euler went on to write a complete book on infinite series, *Introductio in analysin infinitorum*, which was published in 1748.

By concentrating on patterns, rather than arithmetic, mathematicians were thus able to handle the infinite. The most significant consequence of the study of infinite patterns took place in the second half of the seventeenth century. Working independently, Isaac Newton, in England, and Gottfried Leibniz, in Germany, developed a phenomenally powerful method to handle continuous change. Their achievement is undoubtedly one of the greatest mathematical feats of all time. By introducing the *differential calculus*, they transformed human life forever. Without the differential calculus, modern technology simply would not exist; there would be no electricity, no telephones, no automobiles, and no heart-bypass surgery. The sciences that led to these—and most other—technological developments depend on the calculus in a fundamental way.

Functions

The differential calculus provides a means to describe and analyze motion and change; not any motion or change, but motion and change *of a certain kind*. The restriction here is that you have to be presented with a *pattern* that describes the motion or change. For, in concrete terms, the differential calculus is a collection of techniques for *the manipulation of patterns*. (The word 'calculus' is a Latin word that means 'pebble'. The use of pebbles in early counting systems led to the adoption of the word 'calculus' to mean a method for computing.)

The basic operation of the differential calculus is the process known as differentiation. The aim of differentiation is to obtain the rate of change of some changing quantity. In order to do this, the 'value' or 'position' or 'path' of that quantity has to be given by means of an appropriate formula. Differentiation then acts upon that formula, to produce another formula that gives the rate of change. Thus, *differentiation is a process that turns formulas into other formulas.*

For example, suppose that a car travels along a road, and that the distance traveled along the road, say x, varies with the time, t, according to the formula

$$x = 5t^2 + 3t.$$

Then, according to the differential calculus, the speed s (i.e. the rate of change of position) at any time t is given by the formula

$$s = 10t + 3.$$

The formula $10t + 3$ is the result of differentiating the formula $5t^2 + 3t$. (You will see shortly just how differentiation works in this case.)

Notice that the speed of the car is not a constant in this example; the speed varies with time, just as does the distance. Indeed, the process of differentiation may be applied a second time to obtain the acceleration (i.e. the rate of change of the speed). Differentiating the formula $10t + 3$ produces the acceleration

$$a = 10,$$

which in this case is a constant.

The fundamental mathematical object to which the process of differentiation applies is the *function*. Without the notion of a function, there can be no calculus. Just as arithmetical addition is an operation that is performed on numbers, so differentiation is an operation that is performed on functions. But what exactly is a function? The simplest answer is that, in mathematics, a function is a rule that, given one number, allows you to calculate another. (Strictly speaking, this is a special case, but it is adequate for understanding how the calculus works.)

For example, a polynomial formula such as

$$y = 5x^3 - 10x^2 + 6x + 1$$

determines a function. Given any value for x, the formula tells you how to compute a corresponding value for y. For instance, given the value $x = 2$, you may compute

$$y = 5 \times 2^3 - 10 \times 2^2 + 6 \times 2 + 1$$
$$= 40 - 40 + 12 + 1 = 13.$$

Other examples are the trigonometric functions, $y = \sin x$, $y = \cos x$, $y = \tan x$. For these functions there is no simple way to compute the value of y as in the case of a polynomial. Their familiar definitions are given in terms of ratios of the various sides of right-angled triangles, but those definitions apply only when the given x is an angle less than a right angle. The mathematician defines the tangent function in terms of the sine and cosine functions as

$$\tan x = \frac{\sin x}{\cos x},$$

and defines the sine and cosine functions by means of infinite series:

$$\sin x = x - \frac{x^3}{3!} + \frac{x^5}{5!} - \frac{x^7}{7!} + \cdots,$$

$$\cos x = 1 - \frac{x^2}{2!} + \frac{x^4}{4!} - \frac{x^6}{6!} + \cdots.$$

For any number n, $n!$ (read "n-factorial") is equal to the product of all numbers from 1 to n inclusive. Thus $3! = 1 \times 2 \times 3 = 6$.

The series for $\sin x$ and $\cos x$ always give a finite value, and may be manipulated more or less like finite polynomials. These series give the usual values when x is an angle of a right-angled triangle; their advantage is that they give a value for any number x.

Still another example of a function is the exponential function

$$e^x = 1 + \frac{x^1}{1!} + \frac{x^2}{2!} + \frac{x^3}{3!} + \cdots.$$

Again, this infinite series always gives a finite value, and may be manipulated like a finite polynomial. Putting $x = 1$, you get

$$e = e^1 = 1 + \frac{1}{1!} + \frac{1}{2!} + \frac{1}{3!} + \cdots .$$

The mathematical constant e that is the value of this infinite series is an irrational number. Its decimal expansion begins 2.71828.

Computing Gradients

Algebraic formulas such as polynomials or the infinite series for the trigonometric or exponential functions are a very precise, and extremely useful, way to describe a certain kind of abstract pattern. The pattern in these cases is a pattern of association between pairs of numbers: the independent variable or argument, x, that you start with, and the dependent variable or value, y, that results. In many cases, this pattern can be illustrated by means of a graph, as illustrated in the figure on this page. The graph of a function shows at a glance how the variable y is related to the variable x.

For example, in the case of the sine function, as x increases from 0, y also increases, until somewhere near $x = 1.5$ (the exact point is $x = \pi/2$) y starts to decrease; y becomes negative around $x = 3.1$ (precisely, when $x = \pi$), continues to decrease until around $x = 4.7$ (precisely, $x = 3\pi/2$), then starts to increase again.

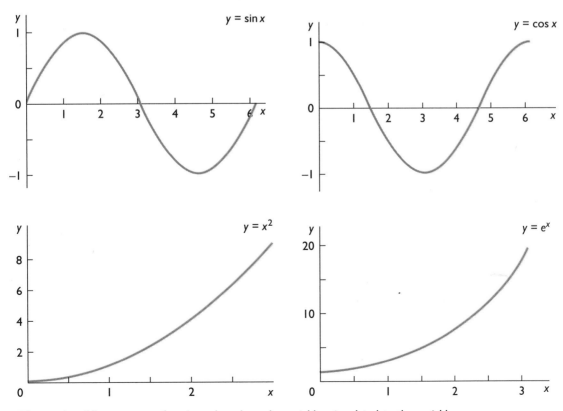

The graphs of four common functions show how the variable y is related to the variable x.

The task facing Newton and Leibniz was this: how do you find the rate of change of such a function, that is, the rate of change of y with respect to x? In terms of the graph, this is the same as finding the gradient of the curve—how steep is it? The difficulty is that the gradient is not constant; at some points the curve is climbing fairly steeply (large, positive gradient), at other points the curve is almost horizontal (gradient close to zero), and at still other points the curve is falling fairly steeply (large, negative gradient).

In summary, just as the value of y depends on the value of x, so too the gradient at any point depends on the value of x. In other words, the gradient of a function *is itself a function*, a second function. The question now is, given a formula for a function—that is to say, a formula that describes the pattern relating x to y—can you find a formula that describes the pattern relating x to the gradient?

The method that both Newton and Leibniz came up with is, in essence, as follows. For simplicity, consider the function $y = x^2$, whose graph is shown below. As x increases, not only does y increase, but the gradient also increases. That is, as x increases, not only does the curve climb higher, it also gets steeper. Given any value for x, the height of the curve for that value of x is given by computing x^2, but what do you do to x in order to compute the gradient for that value of x?

The idea is this. Look at a second point a short distance h to the right of x. Referring to the graph, the height of the point P on the curve is x^2, and the height of Q is $(x + h)^2$. The curve bends up as you go from P to Q, but if h is fairly small (as shown), the difference between the curve and the straight line joining P to Q is also small. So the gradient of the curve at P will be close in value to the gradient of this straight line.

The point of this move is that it is easy to compute the gradient of a straight line: you just divide the increase in the height by the increase in the horizontal direction. In this case, the increase in the height is $(x + h)^2 - x^2$ and the increase in the horizontal direction is h, so the gradient of the straight line from P to Q is

$$\frac{(x + h)^2 - x^2}{h}.$$

Using elementary algebra, the numerator in this fraction simplifies as follows:

$$(x + h)^2 - x^2 = x^2 + 2xh + h^2 - x^2 = 2xh + h^2.$$

So the gradient of the straight line PQ is

$$\frac{2xh + h^2}{h}.$$

Canceling h from this fraction leaves

$$2x + h.$$

Now you have a formula for the gradient of the straight line from P to Q. But what about the gradient of the curve $y = x^2$ at the point P, which is what you started out trying to compute? This is where both Newton and Leibniz made their brilliant and decisive move. They argued as follows. Replace the static situation with a dynamic one, and think about what happens when the distance h that separates the two points P and Q along the x-direction is made smaller and smaller.

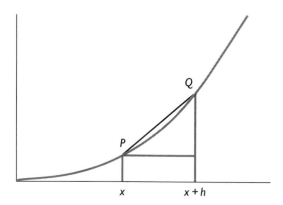

The derivative for the function $y = x^2$.

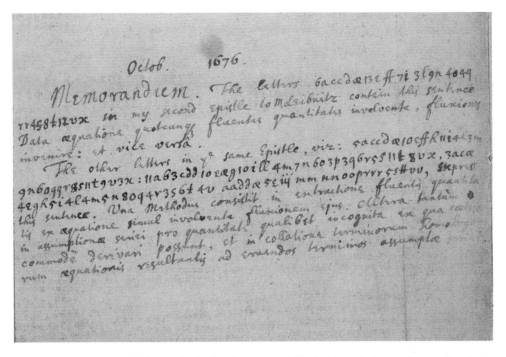

A page from Newton's *Waste Book,* dated October 1676. Written in Newton's own hand, this passage records details of his correspondence with Leibniz, in which Newton explained some of his methods of the calculus.

As h gets smaller, the point Q moves closer and closer to the point P, and, for each value of h, the formula $2x + h$ gives the corresponding value for the gradient of the straight line PQ. For instance, if you take $x = 5$, and let h successively assume each of the values 0.1, 0.01, 0.001, 0.0001, and so on, then the corresponding PQ gradients are 10.1, 10.01, 10.001, 10.0001, ... And there at once you see an obvious numerical pattern: the PQ gradients are approaching the *limiting value* 10.0.

But by looking at the diagram and picturing this process geometrically, another pattern can be observed, a geometric pattern: as h gets smaller, and Q approaches P, the difference between the gradient of the straight line PQ and the gradient of the curve at P also gets smaller, and indeed the limiting value of the gradient of PQ will be the gradient of the curve at P.

So, for the point $x = 5$, the gradient of the curve at P will be 10.0. More generally, the gradient of the curve at P for an arbitrary point x will be $2x$. That is to say, the gradient of the curve at x is given by the formula $2x$.

For the record, Newton's approach was not exactly like this. His main concern was physics; in particular, he was interested in planetary motion. Instead of having a variable y varying with a variable x in a geometric fashion as represented on a graph, Newton thought of a distance r (for radius) varying with time t, say $r = t^2$. He referred to the given function as the *fluent* and the gradient function as the *fluxion*, which he denoted by $\dot{r}$. (So, if the fluent is $r = t^2$, the fluxion will be $2t$.) Obviously, $\dot{r}$ will be some form of speed or velocity (i.e. a rate of change of distance). For the small increment that I earlier denoted by h, Newton used the symbol o, to

The Fathers of the Calculus

Isaac Newton, whose portrait is shown at the right, was born on Christmas Day, 1642, in the Lincolnshire village of Woolsthorpe. In 1661, following a fairly normal grammar school education, he entered Trinity College, Cambridge, where, largely through self study, he acquired a mastery of astronomy and mathematics. In 1664 he was promoted to a 'scholar', a status that provided him with four years of financial support towards a master's degree.

It was on his return home to Woolsthorpe in 1665, when the university was forced to close because of the bubonic plague, that the twenty-three-year-old Newton embarked upon one of the most productive two years of original scientific thought the world has ever seen. The invention of the method of fluxions (the differential calculus) and the inverse method of fluxions (the integral calculus) were just two of a flood of accomplishments in mathematics and physics that he made during the years 1665 and 1666.

In 1668, Newton completed his master's degree and was elected a Fellow of Trinity College, a lifetime position. A year later, when Isaac Barrow resigned his prestigious Lucasian Chair of Mathematics in order to become Chaplain to the King, Newton was appointed to the position.

An overwhelming fear of criticism kept Newton from publishing a great deal of his work, including the calculus, but in 1684 the astronomer Edmund Halley persuaded him to prepare for publication some of his work on the laws of motion and gravitation. The eventual appearance, in 1687, of *Philosophiae Naturalis Principia Mathematica* was to change physical science for all time, and established Newton's reputation as one of the most brilliant scientists the world had—and has—ever seen.

In 1696, Newton resigned his Cambridge chair to become Warden of the Royal Mint. It was while he was in charge of the British coinage that he published, in 1704, his book *Opticks*, a mammoth work outlining the optical theories he had been working on during his Cambridge days. In an appendix to this book he gave a brief account of the method of fluxions that he had developed forty years previously. It was the first time he had published any of this work. A more thorough account, *De Analysi*, had been in private circulation among the British mathematical community from the early 1670s onward, but was not published until 1711. A complete account of the calculus written by Newton did not appear until 1736, nine years after his death.

Just prior to the appearance of *Opticks*, Newton was elected President of the Royal Society, the ultimate scientific accolade in Great Britain, and, in 1705, Queen Anne bestowed on him a knighthood, the ultimate royal tribute. The once shy, frail young child from a small Lincolnshire village was to spend the remaining years of his life regarded as little less than a national treasure.

Sir Isaac Newton died in 1727 at the age of 84, and was buried in Westminster Abbey. His epitaph in the Abbey reads: "Mortals, congratulate yourselves that so great a man has lived for the honor of the human race."

Born in 1646, Gottfried Wilhelm Leibniz was a child prodigy, who made good use of the sizable scholarly library of his father, a professor of philosophy. By the time he was fifteen years old, the young Leibniz was ready to enter the University of Leibzig. Five years later, he had completed his doctorate, and was set to embark on an academic career, when he decided to leave university life and enter government service.

In 1672, Leibniz became a high-level diplomat in Paris, from where he made a number of trips to Holland and Britain. These visits brought him into contact with many of the leading academics of the day, among them the Dutch scientist Christian Huygens, who inspired the young German diplomat to take up once more his studies in mathematics. It proved to be a fortuitous meeting, for, by 1676, Leibniz had progressed from being a virtual novice in mathematics to having discovered for himself the fundamental principles of the calculus.

Or had he? When Leibniz first published his findings in 1684, in a paper in the journal *Acta Eruditorum*, for which he was the editor, many of the British mathematicians of the time cried foul, accusing Leibniz of taking his ideas from Newton. Certainly, on a visit to the Royal Society in London in 1673, Leibniz had seen some of Newton's unpublished work, and, in 1676, in response to a request for further information about his discoveries, Newton had written two letters to his German counterpart, providing some of the details.

Though the two men themselves largely stayed out of the debate, the argument between the British and German mathematicians over who had invented the calculus grew heated at times. Certainly, Newton's work had been carried out before Leibniz's, but the Englishman had not published any of it. In contrast, not only had Leibniz published his work promptly, but his more geometric approach led to a treatment that is in many ways more natural, and which quickly caught on in Europe. Indeed, to this day, Leibniz's geometric approach to differentiation is the one generally adopted in calculus classes the world over, and Leibniz's nota-tion (dy/dx) for the derivative is in widespread use, whereas Newton's approach in terms of physical motion and his notation $\dot{r}$ are rarely used outside of physics.

Today, the general opinion is that, although Leibniz clearly obtained some of his ideas from reading part of Newton's work, the German's contribution was undoubtedly significant enough to grant both men the title of 'father of the calculus'.

Like Newton, Leibniz was not content to spend his entire life working in mathematics. He studied philosophy, he developed a theory of formal logic, a forerunner of today's 'symbolic logic', and he became an expert in the Sanskrit language and the culture of China. In 1700, he was a major force in the creation of the Berlin Academy, of which he was president until his death in 1716.

In contrast to Newton, who was given a state funeral in Westminster Abbey, Germany's creator of the calculus was buried in quiet obscurity. His portrait appears below.

suggest a quantity that was close to, but not quite equal to, 0.

Leibniz, on the other hand, approached the issue as a geometric problem of finding gradients of x, y curves, which is the approach I adopted above. He used the notation dx in place of my h, and dy to denote the corresponding, small difference in y values (the difference in height between P and Q). He denoted the gradient function by $\frac{dy}{dx}$, a notation obviously suggestive of a ratio of two small increments. (The notation dx is normally read 'dee-ex', dy is read 'dee-wye', and $\frac{dy}{dx}$ is read 'dee-wye by dee-ex'.)

For both men, however, the important starting point was to have a functional relation linking the two quantities, either

$$r = \text{some formula involving } t$$

in Newton's case or

$$y = \text{some formula involving } x$$

for Leibniz. In modern terminology, we say that r is *a function of* t, and use notations such as $r = f(t)$ or $r = g(t)$, and analogously in the x, y version.

Motivation and notation aside, the crucial step made by both Newton and Leibniz was to shift attention from the essentially *static* situation concerning the gradient at a particular point P to the *dynamic* process of successive approximation of the gradient by gradients of straight lines starting from P. It was by observing numerical and geometric patterns *in this process of approximation* that Newton and Leibniz were able to arrive at the right answer.

Moreover, their approach works for a great many functions, not just for the simple example considered above. For example, if you start with the function x^3, you get the gradient function $3x^2$, and, more generally, if you start with the function x^n, where n is any natural number, the gradient function works out to be nx^{n-1}. With this, you have another easily recognizable, if somewhat unfamiliar, pattern, the pattern that takes xn to nx^{n-1} for any value of n. This is a pattern of differentiation.

It should be stressed that what Newton and Leibniz were doing was not at all the same as setting the value of h *equal* to 0. True enough, in the case of the very simple example above where the function is x^2, if you simply set $h = 0$ in the gradient formula $2x + h$, then you obtain $2x$, which is the right answer. But if $h = 0$, then the points Q and P are one and the same, so there is no straight line PQ. Remember that, though a factor of h was cancelled to obtain a simplified expression for the gradient of PQ, this gradient is the ratio of the two quantities $2xh + h^2$ and h, and if you put $h = 0$ then this ratio reduces to the division of 0 by 0, which is meaningless.

This point was a source of considerable misunderstanding and confusion both at the time Newton and Leibniz were doing their work and for many subsequent generations. To modern mathematicians, used to regarding mathematics as the science of patterns, the idea of looking for numerical and geometric patterns in a process of successive approximation is not at all strange, but, back in the seventeenth century, even Newton and Leibniz could not formulate their ideas with sufficient precision to silence their many critics. The most notable of those critics was the English philosopher Bishop Berkeley, who, in 1734, published a stinging critique of the calculus.

Leibniz struggled to make himself clear with talk of his dx and dy being "infinitely small" and "indefinitely small" quantities, and when he failed to come up with a sound argument to support his manipulations of these entities, he wrote, "[you may] think that such things are utterly impossible; it will be sufficient simply to make use of them as a tool that has advantages for the purpose of calculation."

While not going so far as to mention the 'infinitely small', Newton referred to his fluxion as the "ultimate ratio of evanescent increments," to which Berkeley retorted in his 1734 critique, "And what are these fluxions? The velocities of evanescent increments. And what are these same evanescent increments? They are neither finite quantities, nor quantities infinitely small, nor yet nothing. May we not call them ghosts of departed quantities?"

If you think of what Newton and Leibniz were doing in a static fashion, with the quantity h a small but fixed quantity, then Berkeley's objections are entirely correct. But if you regard h as a variable, and concentrate not on the given function but rather on the process of approximation that arises when h approaches 0, then Berkeley's argument can no longer be sustained.

To construct a reliable defense against Berkeley, you have to work out a rigorous mathematical theory of approximation processes, which neither Newton nor Leibniz was able to do. Indeed, it was not until 1821 that the Frenchman Augustin Louis Cauchy developed the key idea of a *limit* (of a varying quantity), and it was a few years later still that the German Karl Weierstrass provided a formal definition of the notion of a limit. Only then was the calculus placed on a sound footing, almost two hundred years after its invention.

Augustin Louis Cauchy (1789–1857).

Why did the formulation of a rigorous theory take so long, and, more intriguingly, how was it possible to develop such a powerful, and reliable, tool as the calculus without being able to provide a logical explanation of why it worked?

The method worked because the intuition that drove both Newton and Leibniz was a sound one: they knew that they were working with a dynamic process of successive approximation. Indeed, in his book *Principia*, Newton came very close to achieving the correct formulation with the explanation: "Ultimate ratios in which quantities vanish are not, strictly speaking, ratios of ultimate quantities, but limits to which the ratios of these quantities, decreasing without limit, approach." In other words, to find the gradient function for, say, x^2, it is permissible to determine what happens to the ratio $(2xh + h^2)/h$ as h approaches 0, but you may not set $h = 0$. (Note that cancellation of h to give the expression $2x + h$ for the ratio is only permissible if $h \neq 0$.)

But neither Newton nor Leibniz, nor anyone else until Cauchy and Weierstrass, was able to capture the notion of a limit in a precise mathematical fashion. And the reason is that they were not able to 'step back' quite far enough to discern the appropriate pattern in a static way. For remember, the patterns captured by mathematics are static things, even if they are patterns of motion. Thus, if Newton were thinking about, say, the motion of a planet whose position varies according to the square of the time, he would capture this dynamic situation by means of the static formula x^2—'static' because it simply represents a *relationship* between pairs of numbers. The dynamic motion is captured by a static function.

The key to putting the differential calculus on a rigorous footing was to observe that the same idea can be applied to the dynamic process of approximation to the gradient. The dynamic process of obtaining closer and closer approximations to the gradient as the increment h approaches zero can also be captured in a static fashion, in terms of a function of h. Here is how Weierstrass did just that.

Suppose you have some function $f(h)$; in the example we have been looking at, $f(h)$ would be the quotient $(2xh + h^2)/h$ (where x is regarded as fixed, h as the variable). Then, to say that a number l (which will be $2x$ in our example) is the *limit* of the function $f(h)$ as h approaches 0 means, precisely:

For any $\epsilon > 0$, there exists a $\delta > 0$ so that, if $0 < |h| < \delta$, then $|f(h) - l| < \epsilon$.

Unless you have seen this statement before, you are unlikely to be able to fathom out its meaning. It did, after all, take some two hundred years for mathematicians to arrive at this definition. The point to notice, though, is that it says nothing about any kind of (dynamic) process; it simply refers to the *existence* of numbers δ having a certain property. In this respect, it is just the same as the original step, which Newton performed all the time, of capturing motion by means of a formula. By denoting time by a (static) variable t, Newton could capture motion by means of a formula involving t. By likewise treating h as a variable, Weierstrass was able to capture the notion of a limit (of a sequence of approximations) by means of a formal definition involving h. Newton captured the t-pattern, Weierstrass captured the h-pattern.

Incidentally, though Cauchy developed an extensive theory of limits, adequate for the calculus, he was still working with the notion of a dynamic approximation process. Thus, he only put the calculus 'on a firm footing' in the sense that he reduced the problem to that of providing a precise definition of limit. That final, key step was performed by Weierstrass. But why wasn't Newton or Leibniz or even Cauchy able to do this? After all, each of these great mathematicians was well accustomed to the use of variables to capture motion and the use of formulas to capture patterns of motion. Almost certainly, at issue is the level of process that the human mind can cope with as an entity in itself. At the time of Newton and Leibniz, to regard a function as an *entity*, rather than as a process of change or motion, was already a significant cognitive accomplishment. To subsequently regard the process of successively approximating the gradient of that function as yet another entity in its own right was simply too much. Only after the passage of a suitably long period of time, and growing familiarity with the techniques of the calculus, could anyone make this second conceptual leap. Great mathematicians can perform amazing feats, but they are still only human. Cognitive advances take time, often many generations.

Since their intuitions concerning approximation (or limit) processes were so good, Newton and Leibniz were able to develop their 'differential calculus' into a reliable and extremely powerful tool even though they could not fully understand why the tool worked. To do this, they regarded functions as mathematical objects to be studied and manipulated, and not just as recipes for computations. They were guided by the various patterns that arise in computing successive approximations to gradients (and

Karl Weierstrass (1815–1897).

Fourier Analysis

Today's music synthesizer, which uses computer technology to create complex sounds from the pure notes produced by simple oscillator circuits, is a direct consequence of both the calculus and the development of techniques to manipulate infinite series. Though the technology is very recent, the mathematics behind it was worked out as long ago as the late eighteenth century, by Jean d'Alembert, Daniel Bernoulli, Leonhard Euler, and Joseph Fourier. It is known as 'Fourier analysis'. Fourier analysis deals not with infinite series of numbers but infinite series of functions.

A striking application of the theory is that, given enough tuning forks, you could give a performance of Beethoven's Ninth Symphony, complete in every way, including the choral part. At least, in principle you could; in practice, it would take a very large number of tuning forks indeed to create the complex sounds normally produced by brass, woodwinds, strings, percussion, and human voices. But, in principle, it can be done.

The crux of the matter is that any sound wave, such as the one shown in the top figure below, or indeed any wave of any kind, can be obtained by adding together an infinite series of sine waves, pure waveforms, as shown in the bottom figure below. (A tuning fork produces a sound whose waveform is a sine wave.) For example, the three sine waves in the figure at the top of the page add together to give the more complex wave shown beneath them. Of course, this is a particularly simple example. In practice, it may take a great many individual sine waves to give a particular waveform; mathematically, it may require an infinite number.

The precise statement is known as Fourier's theorem. It gives a mathematical description of any phenomenon, such as a sound wave, that can be con-

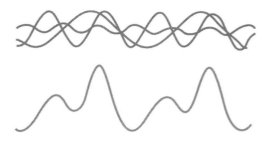

The three waves at the top sum to form the wave below.

sidered as a 'periodic' function of time, that is, a function that keeps on repeating some cycle of values. The theorem says that if y is such a periodic function of time, and if the frequency with which y cycles through its period is, say, 100 times a second, then y can be expressed in a form such as

$$y = 4 \sin 200\pi t + 0.1 \sin 400\pi t \\ + 0.3 \sin 600\pi t + \cdots .$$

This sum may be finite, or it may continue indefinitely to give an infinite series. In each term, the time t is multiplied by 2π times the frequency. The first term is called the first harmonic and its frequency is called the fundamental frequency (100 in the example given). The other terms are called higher harmonics, and all have frequencies that are exact multiples of the fundamental frequency. The coefficients (4, 0.1, 0.3, and so on) must all be adjusted to give the particular waveform y. Determination of these coefficients, from observed values of the function y, constitutes what is known as the Fourier analysis of y. It uses various techniques from the calculus.

In essence, Fourier's theorem tells us that the pattern of any sound wave, or indeed of any kind of wave, no matter how complex, can be built up from the simple, pure wave pattern produced by the sine function. Interestingly, Fourier did not prove this result. What he did was formulate the theorem and give a 'plausibility argument', using some very dubious reasoning that would certainly not be accepted as valid today. However, implicitly acknowledging that the most significant step was to identify the pattern, the mathematical establishment has always credited Fourier with the result.

A typical sound wave.

A sine wave.

other changing quantities) associated with those functions; but they were not able to step back and regard those patterns of approximations as being themselves objects for mathematical study.

The Differential Calculus

The process of going from the formula for a curve to a formula for the gradient of that curve is known as differentiation. (The name reflects the idea of taking small 'differences' in the x and y directions and computing the gradients of the resulting straight lines.) The gradient function is called the derivative of the initial function (from which it is 'derived'). So, in the example we have been looking at, the function $2x$ is the derivative of the function x^2. Similarly, the derivative of the function x^3 is $3x^2$ and, in general, the derivative of the function x^n for any natural number n is nx^{n-1}.

The power of Newton's and Leibniz's invention was that the number of functions that could be differentiated was greatly enlarged by the development of a *calculus*, a series of rules for differentiating complicated functions. The development of this calculus also accounts for the method's enormous success in different applications, despite the dependence on methods of reasoning that were not fully understood. People knew *what* to do, even if they did not know why it worked. Many of the students in today's calculus classes have the same experience.

The rules of the calculus are more conveniently described using modern terminology, whereby arbitrary functions of x are denoted by expressions such as $f(x)$ or $g(x)$, and their derivatives (which are also functions of x) are denoted by $f'(x)$ and $g'(x)$, respectively. So, for example, if $f(x)$ is used to denote the function x^5, then $f'(x) = 5x^{5-1} = 5x^4$.

One of the rules of the calculus gives the derivative of the function $Af(x)$ (i.e. $A \times f(x)$), when A is any fixed number (i.e. a constant). The derivative is simply A times the derivative of $f(x)$, or the function $Af'(x)$. For example, the deriva-

tive of the function $41x^2$ is $41 \times 2x$, which simplifies to $82x$.

Another rule is that the derivative of a 'sum function' of the form $f(x) + g(x)$ is simply the sum of the derivatives of the individual functions, namely $f'(x) + g'(x)$. So, for example, the derivative of the function $x^3 + x^2$ is $3x^2 + 2x$. A similar rule applies for 'difference functions' $f(x) - g(x)$.

Using the above three rules, it is possible to differentiate any polynomial function, since polynomials are built up from powers of x using constant multiples, addition, and subtraction. For example, the derivative of the function $5x^6 - 8x^5 + x^2 + 6x$ is $30x^5 - 40x^4 + 2x + 6$.

In this last example, notice what happens when you differentiate the function $6x$. The derivative is 6 times the derivative of the function x. Applying the rule that takes a power x^n to the derivative nx^{n-1}, the derivative of the function x, which is the same as x^1, is $1x^{1-1}$, which is $1x^0$. But any number raised to the power 0 is 1. So the derivative of the function x is just 1.

What happens when you try to differentiate a fixed number, say 11? This problem would arise if you tried to differentiate the polynomial $x^3 - 6x^2 - 4x + 11$. Remember that differentiation is a process that applies to formulas, not to numbers; it is a method to determine gradients. So in order to differentiate 11, you have to think of it not as a number but as a function, the function that gives the value of 11 for any value of x. Thinking of a number in this way may seem strange, but when drawn as a graph it is perfectly natural. The 'function' 11 is just a horizontal line drawn 11 units above the x-axis, that is, drawn through the point 11 on the y-axis, as in the graph on the facing page. You don't need calculus to figure out the gradient of this function; it is 0. In other words, the derivative of a 'constant function', such as the function 11, is 0.

It was in order to provide a foundation for rules of the calculus such as the ones just described that Cauchy developed his theory of limits. In both the case of multiplication by a fixed number and the case of the sum or difference of two functions, the rules

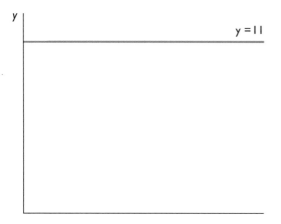

The constant function y = 11.

(or patterns) of differentiation turn out to be very straightforward. In the case of multiplication of one function by another, the pattern is a little more complicated. The formula for the derivative of a function of the form $f(x)g(x)$ is

$$f(x)g'(x) + g(x)f'(x).$$ *why?*

For instance, the derivative of the function $(x^2 + 3)(2x^3 - x^2)$ is

$$(x^2 + 3)(6x^2 - 2x) + (2x^3 - x^2)(2x + 0).$$

Other functions for which the derivatives have simple patterns are the trigonometric functions: the derivative of $\sin x$ is $\cos x$, the derivative of $\cos x$ is $-\sin x$, and the derivative of $\tan x$ is $1/(\cos x)^2$.

Even simpler is the pattern for the exponential function: the derivative of e^x is just e^x itself, which means that the exponential function has the unique property that its gradient at any point is exactly equal to the value at that point.

In the cases of the sine, cosine, and exponential functions (though not for all functions defined by infinite series), it turns out that the derivative can be obtained by differentiating the infinite series term by term, as if it were a finite polynomial. If you do this, you will be able to verify the above differentiation results for yourself.

Differentiation of the natural logarithm function $\log x$ also produces a simple pattern: the derivative of $\log x$ is the function $1/x$.

Differential Equations

In 1986, at Chernobyl in the Ukraine, a disaster at a nuclear power plant caused the release of radioactive material into the atmosphere. The authorities claimed that the amount of radioactivity in the surrounding areas would at no stage reach a catastrophically dangerous level for people. How could they arrive at this conclusion? More generally, under such circumstances, how can you predict what the level of radioactivity will be a day or a week in the future, in order that any necessary evacuations or other precautions can be carried out?

The answer is, you solve a *differential equation*. You want to know the amount of radioactivity in the atmosphere at any time t after the accident. Since the radioactivity varies over time, it makes sense to write it as a function of time, $M(t)$, but, when you start the investigation, you probably do not have a formula with which to calculate the value at any given time. However, physical theory leads to an equation that connects the rate of growth of radioactive material, $\dfrac{dM}{dt}$, with the constant rate, k, at which the radioactive material is released into the atmosphere and the constant rate, r, at which the radioactive material decays. The equation is

$$\frac{dM}{dt} = r\left(\frac{k}{r} - M\right).$$

This is an example of a differential equation, an equation that involves one or more derivatives. 'Solving' such an equation means finding a formula for the unknown function, $M(t)$. Depending on the equation, this may or may not be possible. In the case of the radioactive contamination scenario just described, the equation is particularly simple, and a solution can be obtained. It is the function

$$M(t) = \frac{k}{r}(1 - e^{-rt}).$$

When you draw a graph of this function, the first of the four shown on this page, you see that it grows rapidly at first, but then gradually levels off, getting closer to, but never reaching, the limiting value k/r. Thus the highest level the contamination will ever reach is no greater than k/r.

The same kind of differential equation arises in physics, for example in Newton's law of cooling; in psychology, as a result of studies of learning (the so-called Hullian learning curve); in medicine, where it describes the rate of intravenous infusion of medication; in sociology, in measurement of the spread of information by mass media; and in economics, for example in the phenomena of depreciation, sales of new products, and growth of a business. The overall pattern is one of 'limited growth', where some quantity is growing toward a maximum value.

In general, a differential equation arises whenever you have a quantity that is subject to change, and where theory provides you with a growth pattern in the form of an equation. Strictly speaking, the changing quantity should be one that changes continuously, which means that it can be captured by means of a function of a real-number variable. However, change in many real-life situations consists of a large number of individual, discrete changes that are miniscule compared with the overall scale of the problem, and in such cases there is

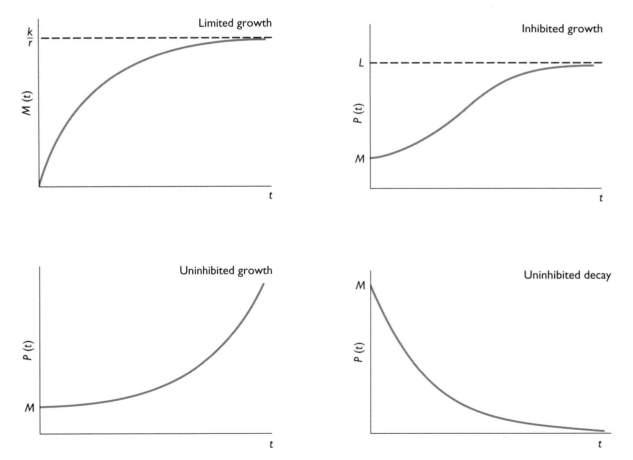

The four graphs are of solutions to differential equations of four different forms.

Calculus can be applied to the study of many real-life phenomena. For example, differential equations were solved to create this computer simulation of airflow over the body of an F-18 fighter plane.

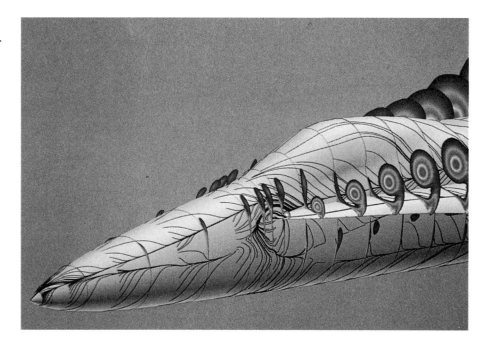

no harm in simply assuming that the whole changes continuously. This enables the full power of the calculus to be brought to bear in order to solve the differential equation that results. Most applications of differential equations in economics are of this nature: the actual changes brought about in an economy by single individuals and small companies are so small compared to the whole, and there are so many of them, that the whole system behaves as if it were experiencing continuous change.

Other kinds of change give rise to differential equations of other forms. For example, the differential equation

$$\frac{dP}{dt} = rP$$

describes 'uninhibited growth', where $P(t)$ is the size of some population and r is a fixed growth rate. The solution function in this case is

$$P(t) = Me^{rt},$$

where M is the initial size of the population. The graph of this solution is shown on the facing page at the lower left. Over the short term, animal populations, epidemics, and cancers can grow according to this pattern, as can inflation.

Over longer periods, a far more likely scenario than uninhibited growth is that of 'inhibited growth', which is captured by the differential equation

$$\frac{dP}{dt} = rP(L - P),$$

where L is some limiting value for the population. This equation has the solution function

$$P = \frac{ML}{M + (L - M)e^{-Lrt}}.$$

If you draw the graph of such a function, as in the graph on the facing page at top right, you see that it begins at the initial value M, grows slowly at first, then starts to grow more rapidly until it nears the limiting value L, when the rate of growth steadily slows down.

Finally, the differential equation

$$\frac{dP}{dt} = -rP$$

describes 'uninhibited decay'. The solution function is

$$P(t) = Me^{-rt}.$$

Radioactive decay and the depletion of certain natural resources follow this pattern, which is also graphed on page 92.

More complex forms of differential equation often involve the 'second derivative', the derivative of the derivative. This is particularly true of many differential equations that arise in physics.

The task of finding solutions to differential equations is an entire branch of mathematics in its own right. In many cases, it is not possible to obtain a solution given by a formula; instead, computational methods are used to obtain numerical or graphical solutions.

With differential equations arising in almost every walk of life, their study is a branch of mathematics having enormous consequences for humanity. Indeed, from a quantitative point of view, differential equations describe the very essences of life: growth, development, and decay.

Integration

The development of the differential calculus brought with it a surprising bonus, one that was scarcely to be expected. It turns out that the fundamental patterns of differentiation are the same as the patterns that underlie the computation of areas and volumes. More precisely, the computation of areas and volumes is, essentially, the *inverse* of differentiation—the process of finding gradients. This amazing observation is the basis of an entire second branch of the calculus, the *integral calculus*.

Computing the area of a rectangle or the volume of a cube is a straightforward matter of multiplying the various dimensions, the length, the breadth, the height, and so forth. But how do you compute the area of a figure having curved edges or the volume of a solid with curved surfaces? For instance, what is the area traced out by a parabola, shown shaded in the figure on this page? Or what is the volume of a cone?

The first known attempt to compute the areas and volumes of geometric figures was made by Eudoxus, a student of Plato at the Academy in Athens. Eudoxus invented a powerful, and exceedingly clever, method to compute areas and volumes, known as the 'method of exhaustion'. Using this method, he was able to show that the volume of any cone is equal to one-third the volume of the cylinder having the same base and equal height, a remarkable pattern that is neither obvious nor easy to prove. (See the top figure on the facing page.)

Archimedes used Eudoxus' method to compute the areas and volumes of a number of figures. For example, he found the area traced out by a parabola. This example serves to illustrate how the method of exhaustion works. The idea is to approximate the curve by a series of straight lines, as shown in the figure at the bottom of the facing page. The area traced out by the straight lines consists of two triangles (at the ends) and a number of trapeziums. Since there are simple formulas for the area of a tri-

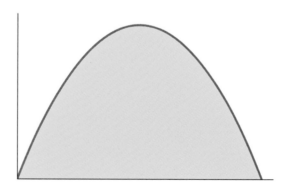

The area traced out by a parabola.

The volume of a cone is equal to one-third the volume of a cylinder having the same base and the same height.

creasing the number of straight lines in the approximation, obtaining better and better approximations to the area beneath the parabola. You stop when you feel the approximation is good enough.

This process is called the 'method of exhaustion' not so much because Eudoxus became exhausted after computing so many approximations, but because the sequence of successive approximating areas would eventually 'exhaust' the entire area beneath the original curve, if continued indefinitely.

Interest in geometric figures such as parabolas and ellipses was revived during the early part of the seventeenth century, when Johannes Kepler observed his now-famous laws of planetary motion, three elegant, and profound, mathematical patterns in the sky: (1) a planet orbits the sun in an ellipse having the sun at one of its two foci; (2) a planet sweeps out equal areas in equal times; and (3) the cube of a planet's distance from the sun is equal to the square of its orbital period.

Mathematicians of the time, among them Galileo, Kepler himself, and, above all, Bonaventura Cavalieri of Italy, computed areas and volumes by means of the method of *indivisibles*. In this approach, a geometric figure is regarded as being made up of an infinite number of 'atoms' of area or volume, which are added together to give the required area or volume. The general idea is illustrated on the following page. Each of the shaded areas is a rectan-

angle and the area of a trapezium, you can compute the area traced out by the straight lines by simply adding together the areas of the triangles and the trapeziums. The answer you get will be an approximation to the area beneath the parabola. You can obtain a *better* approximation by increasing the number of straight lines and repeating the calculation. The method of exhaustion works by gradually in-

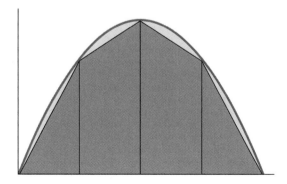

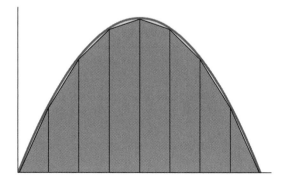

The area beneath the parabola is approximated by the sum of the triangles and trapeziums. The greater the number of subdivisions, the more accurate is the approximation.

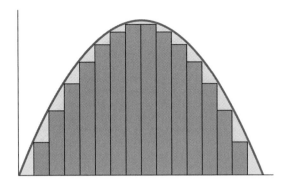

The method of indivisibles.

gle, whose area can be calculated precisely. When there are only a finite number of such rectangles, as shown, adding together all these areas gives an approximation to the area under the parabola. If there were infinitely many rectangles, all infinitesimally wide, the addition would produce the true area—if only it were possible to carry out this infinite computation. In his book *Geometria indivisibilus continuorum*, published in 1635, Cavalieri showed how to handle indivisibles in a reasonably reliable way, in order to come up with the right answers. When placed on a rigorous footing with the aid of the Cauchy–Weierstrass theory of limits, this approach became the modern theory of *integration*.

Eudoxus' method of exhaustion and Cavalieri's method of indivisibles provided means to compute the area or volume of a specific figure. But each method involved a lot of repetitive computation, and you had to start over again from the start each time you were faced with a new figure. In order to provide mathematicians with a versatile and efficient means to compute areas and volumes, more was required: a way to go from a formula for the figure to a formula for the area or volume, much as the differential calculus takes you straight from a formula for a curve to a formula for its gradient.

It was truly amazing not only that there was such a general method, but that it turned out to be a direct consequence of the method of differentiation. As with the differential calculus, the key step

is to consider not the problem of computing a particular area or volume, but the more general task of finding an area or volume *function*.

Take the case of computing areas. The curve shown below traces out an area. More precisely, it determines an area *function*: for any point x, there will be a corresponding area, the area shown shaded in the picture below. (This is a specially chosen, simple case. The general situation is a bit more complicated, but the essential idea is the same.) Let $A(x)$ denote this area, and let $f(x)$ be the formula that determines the original curve. In any particular example, you will know the formula for $f(x)$ but will not have a formula for $A(x)$.

Even if you don't know its formula, $A(x)$ is still a function, and so might have a derivative; in which case, you can ask what its derivative is. And, lo and behold, the answer turns out to be none other than the function $f(x)$, the very formula for the curve that traces out the area in the first place.

This is true for any function $f(x)$ given by a reasonable formula, and can be proved *without knowing a formula for* $A(x)$! The proof depends only on the general *patterns* involved in computing areas and derivatives.

Briefly, the idea is to look at the way the area $A(x)$ changes when you increase x by a small amount h. Referring to the figure on the facing page, the new area $A(x + h)$ can be split up into two parts: $A(x)$ plus a small additional area that is very nearly rectangular, shaded yellow in the figure. The width of this additional rectangle is h; its height is given

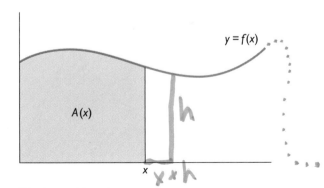

The integral function.

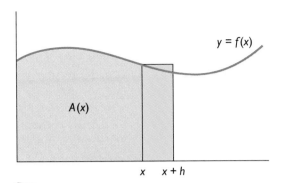

The proof of the fundamental theorem of calculus.

by reading off from the graph: it is $f(x)$. So the area of the extra piece is $h \times f(x)$ (width times height). The entire area is

$$A(x + h) = A(x) + h \times f(x).$$

This equation can be rearranged to look like this:

$$\frac{A(x + h) - A(x)}{h} = f(x).$$

Strictly speaking, this identity is only approximate, since the additional area you have to add to $A(x)$ to give the area $A(x + h)$ is not exactly rectangular. But the smaller h is, the better this approximation will be.

The expression on the left looks familiar, doesn't it? It is precisely the expression that gives the derivative $A'(x)$ when you take the limit as h approaches 0. So, as h becomes smaller and smaller, three things happen: the equation becomes more and more accurate, the expression on the left approaches the limit $A'(x)$, and the right-hand side remains constant at the value $f(x)$, whatever that value is. You finish up with, not an approximation, but the genuine identity

$$A'(x) = f(x).$$

This remarkable result, connecting the tasks of computing gradients and computing areas (it also works for volumes), is known as the *fundamental theorem of the calculus*.

The fundamental theorem of the calculus provides a method to find a formula for $A(x)$. In order to find the area function $A(x)$ for a given curve $f(x)$, you have to find a function whose derivative is $f(x)$. Suppose, for example, that $f(x)$ is the function x^2. The function $\frac{1}{3}x^3$ has the derivative x^2, so the area function $A(x)$ is $\frac{1}{3}x^3$. In particular, if you want to know the area traced out by the curve x^2 up to the point $x = 4$, you set $x = 4$ in this formula to get $\frac{1}{3}4^3$, which works out to be 64/3, or $21\frac{1}{3}$. (Again, I have chosen a particularly simple case that avoids one or two complications that can arise with other examples, but the basic idea is correct.) In order to compute areas and volumes, therefore, you just have to learn how to do differentiation 'backward'. Just as differentiation itself is so routine that it can be programmed into a computer, so too there are computer programs that can perform integration.

The fundamental theorem of the calculus stands as a shining example of the huge gain that can result from a search for deeper, more general, and more abstract patterns. In the cases both of finding gradients and of computing areas and volumes, the ultimate interest may well be in finding a particular number, and yet the key in both cases is to look at the more general, and far more abstract, patterns whereby the gradient and the area or volume change with varying x-values.

The Real Numbers

Whether or not time and space are continuous, or have a discrete, atomic nature, practically all of the scientific and mathematical developments that took place from Greek times up to the end of the nineteenth century were built on the assumption that time and space were continuous and not discrete. Both time and space were regarded as 'continua', which were assumed to avoid the paradoxes of Zeno.

By the time of Newton and Leibniz, the continuum that arose from the physical world of time and space was equated with a continuum of 'real numbers'. Numerical measurements of time and of physical quantities, such as length, temperature, weight, velocity, and so on, were assumed to be 'points' on the continuum. The differential calculus applied to functions involving variables that ranged over the real-number continuum.

In the 1870s, when Cauchy, Weierstrass, Richard Dedekind, and others tried to develop a theory of limits adequate to support the techniques of the calculus, they had to carry out a deep investigation of the nature of the real-number continuum. Their starting point was to regard that continuum as a set of 'points'—the real numbers—arranged in a line that stretches out to infinity in both directions.

The real numbers are an extension of the rational numbers, and many of the axioms for the real numbers are also axioms for the rationals. In particular, this is the case for the arithmetical axioms that specify properties of addition, subtraction, multiplication, and division. The arithmetical axioms ensure that the real numbers are a field (see page 54). There are also axioms that describe the ordering of the real numbers, and again these axioms are also axioms for the rational numbers. The key axiom that distinguishes the real numbers from the rational numbers is the one that allows the development of an adequate theory of limits. Though the rational numbers have all of the necessary arithmetic and order properties needed for the calculus, they are not at all suited to a theory of limits. As formulated by Cauchy, this additional axiom reads as follows:

> Suppose that $a_1, a_2, a_3, \ldots$ is an infinite sequence of real numbers that get closer and closer together (in the sense that, the further along the sequence you go, the closer to 0 the differences between the numbers become). Then there must be a real number, call it l, such that the numbers in the sequence get closer and closer to l (in the sense that, the further along the sequence you go, the closer to 0 the differences between the numbers a_n and the number l become).

The number l is known as the *limit* of the sequence $a_1, a_2, a_3, \ldots$.

Notice that the rational numbers do not have this property. The sequence

$$1, 1.4, 1.41, 1.414, \ldots$$

of successive rational approximations to $\sqrt{2}$ is such that the terms get closer and closer together, but there is no single *rational* number l such that the terms in the sequence get arbitrarily close to l. (The only possibility for the number l is $\sqrt{2}$, which we know is not rational.)

Cauchy's axiom is known as the *completeness axiom*. Cauchy gave a formal construction of the real numbers by starting with the rational numbers and 'adding' to the rational line new points to be the limits of all sequences of rationals that have the property of getting closer and closer together. An alternative construction of the real numbers from the rationals was given by Dedekind.

The construction of the real numbers and the development of rigorous theories of limits, derivatives, and integrals, begun by Cauchy, Dedekind, Weierstrass, and others, was the beginning of the subject nowadays referred to as *real analysis*. These days, a fairly extensive study of real analysis is regarded as an essential component of any college education in mathematics.

Complex Numbers

That what started out as an investigation of motion and change should lead to theories of limits and of the real continuum is perhaps not surprising, given the two-thousand-year-old paradoxes of Zeno. Far more of a surprise is that the invention of the calculus should lead also to the acceptance into the mainstream of mathematics of a system of numbers that includes such a counterintuitive entity as the square root of -1. And yet this is exactly what happened, with Cauchy as one of the principal movers in the development.

The story begins about a hundred years before Newton and Leibniz carried out their work. European mathematicians of the sixteenth century, in particular the Italians Girolamo Cardano and Rafaello Bombelli, began to realize that, in trying to solve algebraic problems, it was sometimes useful to assume the existence of negative numbers, and, moreover, to assume that negative numbers have square roots. Both assumptions were widely regarded as extremely dubious, at worst utter nonsense and at best having solely a utilitarian purpose.

Since the time of the ancient Greeks, mathematicians had known how to manipulate expressions involving minus signs, using rules such as $-(-a) = a$ and $1/-a = -1/a$. However, they felt this was only permissible if the final answer was positive. Their distrust of negative numbers was largely a legacy of the Greek notion of numbers representing lengths and areas, which are always positive. Not until the eighteenth century were negative numbers accepted as bona fide numbers.

It took even longer to accept that the square root of a negative number can be a genuine 'number'. The reluctance to accept these numbers is reflected in the use of the term *imaginary number* for such an entity. As had occurred with negative numbers, mathematicians allowed themselves to manipulate imaginary numbers during the course of a calculation. Indeed, arithmetic expressions involving imaginary numbers can be manipulated using the ordinary rules of algebra. The question was, do such numbers exist?

The question reduces to the existence or otherwise of a single imaginary number, the square root of -1. For, suppose there is such a number as $\sqrt{-1}$. Following Euler, denote it by the letter i, for 'imaginary'. Then the square root of any negative number $-a$ is simply $i\sqrt{a}$, the product of this special number i and the square root of the positive number a.

Ignoring the question whether the number i really exists, mathematicians introduced hybrid numbers of the form $a + bi$, where a and b are real numbers. These are called *complex numbers*. Using the ordinary rules of algebra, together with the fact that $i^2 = -1$, it is possible to add, subtract, multiply, and divide complex numbers; for instance,

$$(2 + 5i) + (3 - 6i) = (2 + 3) + (5 - 6)i$$
$$= 5 - 1i$$
$$= 5 - i$$

and (using a period to denote multiplication)

$$(1 + 2i)(3 + 5i) = 1.3 + 2.5.i^2 + 2.3.i + 1.5.i$$
$$= 3 - 10 + 6i + 5i$$
$$= -7 + 11i.$$

(Division is a little more complicated.)

In present-day terminology, the complex numbers would be said to constitute a field, just as do the rationals and the reals. Instead of being points on a line, as are the rationals and the reals, the complex numbers are points in the plane: the complex number $a + bi$ is the point with coordinates a and b, as shown in the figure below.

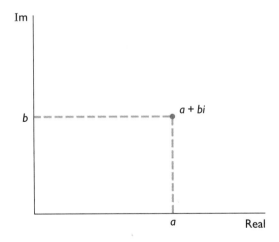

In the complex plane, the horizonal axis is referred to as the 'real axis' and the vertical axis is referred to as the 'imaginary axis', since all real numbers lie on the horizontal axis and all imaginary numbers lie on the vertical axis. All other points in the complex plane represent complex numbers, a sum of a real number and an imaginary number.

The Fundamental Theorem of Algebra

The natural numbers are the most basic number system of all. Though useful for counting, they are not suited to solving equations. Using the natural numbers, it is not possible to solve even such a simple equation as

$$x + 5 = 0.$$

In order to solve equations of this kind, you need to go to the integers.

But the integers are also impoverished, in that they do not allow you to solve a simple linear equation such as

$$2x + 3 = 0.$$

In order to solve equations of this kind, you need to go to the rationals.

The rationals are adequate to solve all linear equations, but do not allow you to solve all quadratic equations; for example, we know that the equation

$$x^2 - 2 = 0$$

cannot be solved in the rationals. The reals are sufficiently rich to solve this quadratic equation. But the reals do not allow you to solve all quadratic equations; for example, you cannot solve the equation

$$x^2 + 1 = 0$$

in the reals. To solve this quadratic equation, you need to go to the complex numbers.

At this point, primed as you are to look for patterns in mathematics, you might be tempted to suppose that this process goes on forever: each time you move to a richer number system, you find yet another kind of equation that you cannot solve. But this is not the case. When you reach the complex numbers, the process comes to a halt. Any polynomial equation

$$a_nx^n + a_{n-1}x^{n-1} + \cdots + a_1x + a_0 = 0,$$

where the coefficients $a_0, \ldots, a_n$ are complex numbers, can be solved in the complex numbers.

This important result is known as the *fundamental theorem of algebra*. It was suspected, but not proved, in the early seventeenth century. Incorrect proofs were supplied by d'Alembert in 1746 and Euler in 1749. The first correct proof was given by Gauss in his 1799 doctoral thesis.

Since the complex numbers are not points on a line, you cannot say which one of two complex numbers is the larger; for the complex numbers there is no such notion. There is, however, a notion of size of sorts. The *absolute value* of a complex number $a + bi$ is the distance from the origin to the number, measured in the complex plane; it is usually denoted by $|a + bi|$. By the Pythagorean theorem:

$$|a + bi| = \sqrt{a^2 + b^2}.$$ why?

The absolute values of two complex numbers may be compared, but it is possible for different complex numbers to have the same absolute value; for example, both $3 + 4i$ and $4 + 3i$ have absolute value 5.

Despite the gradual increase in the use of complex numbers, they were not regarded as bona fide numbers until the middle of the nineteenth century, when Cauchy and others started to extend the methods of the differential and integral calculus to include the complex numbers. Their theory of differentiation and integration of complex functions turned out be so elegant—far more than in the real case—that on aesthetic grounds alone it was impossible to resist the admission of complex numbers as fully paid-up members of the mathematical club.

Euler's Formula

Complex numbers turn out to have connections to many other parts of mathematics. A particularly striking example comes from the work of Euler. In 1748, he discovered the amazing identity

$$e^{ix} = \cos x + i \sin x.$$

This is true for any real number x.

Such a close connection between trigonometric functions, the mathematical constant e, and the square root of -1 is already quite startling. Surely, such an identity cannot be a mere accident; rather, we must be catching a glimpse of a rich, complicated, and highly abstract mathematical pattern that for the most part lies hidden from our view.

In fact, Euler's formula has other surprises in store. If you substitute the value π for x in Euler's formula, then, since $\cos \pi = -1$ and $\sin \pi = 0$, you get the identity

$$e^{i\pi} = -1.$$

Rewriting this as

$$e^{i\pi} + 1 = 0$$

you obtain a simple equation that connects the five most common constants of mathematics: $e, \pi, i, 0$, and 1.

Not the least surprising aspect of the last equation is that the result of raising an irrational number to a power that is an irrational imaginary number can turn out to be a natural number. Indeed, raising an imaginary number to an imaginary power can also give a real-number answer. Setting $x = \pi/2$ in the first equation, and noting that $\cos \pi/2 = 0$ and $\sin \pi/2 = 1$, you get

$$e^{i\pi/2} = i,$$

and, if you raise both sides of this identity to the power i, you obtain (since $i^2 = -1$)

$$e^{-\pi/2} = i^i.$$

Thus, using a calculator to compute the value of $e^{\pi/2}$, you find that

$$i^i = 0.207\ 879\ 576\ 3 \ldots.$$

For, provided it is correct, mathematicians never turn their backs on beautiful mathematics, even if it flies in the face of all their past experience.

In addition to its mathematical beauty, however, complex calculus—or *complex analysis* as it is referred to these days—turned out to have significant applications to, of all things, the theory of the natural numbers. The fact that there was a deep and profound connection between complex analysis and the natural numbers was yet another testimony to the power of mathematical abstraction. The techniques of complex calculus enabled number-theorists to identify and describe number patterns that, in all likelihood, would otherwise have remained hidden forever.

Analytic Number Theory

The first person to use the methods of complex calculus to study properties of the natural numbers—a technique known nowadays as *analytic number theory*—was the German mathematician Bernhard Riemann. In a paper published in 1859, titled *On the number of primes less than a given magnitude*, Riemann used complex calculus to investigate a number-theoretic pattern that had first been observed by Gauss: for large natural numbers N, the number of primes less than N, generally denoted by $\pi(N)$, is approximately equal to the ratio $N/\log N$, where $\log N$ is the natural logarithm of N. (See page 19 for a table of values of the function $\pi(N)$.) Since both $\pi(N)$ and $N/\log N$ grow increasingly large as N increases, you have to formulate this observation with some care. The precise formulation is that the limit of the ratio $\pi(N)/[n/\log N]$ as N approaches infinity is exactly equal to 1. This observation became known as the prime number conjecture.

The closest anyone had come to a proof was a result obtained by Pafnuti Chebyshev in 1852, which said that, for sufficiently large values of N, $\pi(N)/[n/\log N]$ lies between 0.992 and 1.105. In order to obtain this result, Chebyshev had made use of a function introduced by Euler back in 1740, and called the zeta function, after the Greek letter Euler had used to denote this function.

Euler defined the zeta function by the infinite series

$$\zeta(x) = \frac{1}{1^x} + \frac{1}{2^x} + \frac{1}{3^x} + \frac{1}{4^x} + \cdots .$$

The number x here can be any real number greater than 1. If x is less than or equal to 1, this infinite series does not have a finite sum, so $\zeta(x)$ is not defined for such x. If you put $x = 1$, the zeta function gives the harmonic series, considered earlier in the chapter. For any value of x greater than 1, the series yields a finite value.

Euler showed that the zeta function is related to the prime numbers by proving that, for all real numbers x greater than 1, the value of $\zeta(x)$ is equal to the infinite product

$$\frac{1}{1 - (1/2)^x} \times \frac{1}{1 - (1/3)^x} \times \frac{1}{1 - (1/5)^x} \times \cdots ,$$

where the product is over all numbers of the form

$$\frac{1}{1 - (1/p)^x},$$

where p is a prime number.

This connection between the infinite series of the zeta function and the collection of all prime numbers is already quite striking—after all, the primes seem to crop up among the natural numbers in a fairly haphazard way, with little discernible pattern, and yet the infinite series of the zeta function has a very clear pattern, progressing steadily up through all the natural numbers one at a time.

The major step made by Riemann was to show how to extend the definition of the zeta function to give a function $\zeta(z)$ defined on all complex numbers z. (It is customary to denote a complex variable by the letter z, just as x is used to denote a real variable.) To achieve his result, Riemann used a complicated process known as 'analytic continuation'. This process works by extending a certain abstract pattern possessed by Euler's function, though it is a pattern whose abstraction is well beyond the scope of a book such as this.

Why did Riemann go to all this effort? Because he realized that he could prove the prime number conjecture if he was able to understand the complex *zeros* of the zeta function, the solutions to the equation

$$\zeta(z) = 0.$$

Georg Friedrich Bernhard Riemann (1826–1866).

The real zeros of the zeta function are easy to find: they are -2, -4, -6, and so on, the negative even integers. (Remember, Euler's definition of the zeta function in terms of an infinite series only works for real numbers x greater than 1. I am now talking about Riemann's extension of this function.)

In addition to these real zeros, the zeta function also has infinitely many complex zeros. They are all of the form $x + iy$ where x lies between 0 and 1; that is, in the complex plane they all lie between the y-axis and the vertical line $x = 1$. But can we be more precise than this? In his paper, Riemann proposed the hypothesis that all the zeros other than the negative even integers are of the form $\frac{1}{2} + iy$; that is, they lie on the line $x = \frac{1}{2}$ in the complex plane. The prime number conjecture follows from this hypothesis.

Riemann must have based this hypothesis on his understanding of the zeta function and the pattern of its zeros. He certainly had little by way of numerical evidence—that only came much later, with the advent of the computer age. Computations carried out over the last thirty years have showed that the first one and a half billion zeros all lie on the appropriate line. But despite this impressive-looking numerical evidence, the Riemann hypothesis remains unproved to this day. Most mathematicians would agree that it is *the* most significant unsolved problem of mathematics.

The prime number conjecture was finally proved by Jacques Hadamard, and independently by Charles de la Vallée Poussin, in 1896. Their proofs used the zeta function, but did not require Riemann's hypothesis.

Mathematicians had come full circle. Mathematics began with the natural numbers, the building blocks of counting. With the invention of the calculus by Newton and Leibniz, mathematicians tamed the infinite, and were able to study continuous motion. The introduction of the complex numbers and the proof of the fundamental theorem of algebra provided the facility to solve all polynomial equations. Then Cauchy and Riemann showed how to extend the calculus to work for complex functions. Finally, Riemann and others used the resulting theory—a theory of considerable abstraction and complexity—to establish new results about the natural numbers.

I. Rice Pereira,
Oblique Progression (1948)

4

Shape

At first glance, what do you see in the diagram on the following page? Like everyone else, you probably see a triangle. But look more closely, and you will see that there is no triangle *on the page,* merely a collection of three dark-blue disks, with a bite missing out of each. The triangle that you see is an optical illusion, produced without any conscious effort on your part, because your mind and visual system 'fill in' details such as lines and surfaces in order to obtain 'geometric cohesion'. The human visual–cognitive system constantly 'looks for' geometric patterns. In this sense, we are all geometers.

Having acknowledged that we 'see' geometric shapes, what is it that enables you to recognize a triangle as a triangle, be it on the page, made out of wood, or in your mind as in the example shown? Obviously not the size. Nor the color. Nor the thickness of the lines. Rather it is the *shape*. Whenever you see three straight lines joined at their ends to form a closed figure, you recognize that figure as a triangle. You do this because you possess the *abstract concept* of a triangle. Just as the abstract concept of the number 3 tran-

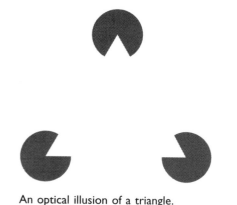

An optical illusion of a triangle.

scends any particular collection of three objects, so too the abstract concept of a triangle transcends any particular triangle. In this sense too, we are all geometers.

We live our lives in large part by being able to recognize, and sometimes ascribe, *shape*. The mathematical study of shape has given rise to several branches of mathematics. The most obvious is *geometry*, which forms the main topic of this chapter. *Symmetry* and *topology*, discussed in the following chapters, study different, in some ways more abstract, patterns of shape.

Euclid's Axioms

The word 'geometry' comes from the Greek: 'geometry' means 'earth measurement'. The mathematical ancestors of today's geometers were the land surveyors of ancient Egypt, who had the task of re-establishing boundaries washed away by the periodic flooding of the Nile; the Egyptian and Babylonian architects, who designed and built temples, tombs, and the obviously 'geometrical' pyramids; and the early seaborne navigators who traded along the Mediterranean coast. Just as these same early civilizations made practical use of numbers without having an obvious concept of 'number', let alone a

theory of such entities, so too their largely utilitarian use of various properties of lines, angles, triangles, circles, and the like was unaccompanied by any detailed, mathematical study.

As mentioned in Chapter 1, it was Thales who, in the sixth century B.C., started the Greek development of geometry as a mathematical discipline—indeed, the first mathematical discipline. Euclid's *Elements,* written around 350 B.C., was largely a book on geometry.

In Book I of *Elements,* Euclid tried to capture the abstract patterns of the *regular* shapes in the plane, namely straight lines, polygons, and circles, by means of a system of definitions and postulates (axioms) for what was to become known as Euclidean geometry. Among the twenty-three initial definitions he set down are:

Definition 1   A point is that which has no part.

Definition 2   A line is a breadthless length. [For Euclid, lines could be either straight or curved.]

Definition 4   A straight line is a line that lies evenly with the points on itself.

Definition 10   When a straight line standing on another straight line makes the adjacent angles equal to one another, each of the equal angles is *right* and the straight line standing on the other is called a *perpendicular* to that on which it stands.

Definition 23   Parallel straight lines are straight lines that, being in the same plane and being produced [i.e. extended] indefinitely in both directions, do not meet one another in either direction.

To the mathematician of today, the first three of the above definitions are unacceptable; they simply replace three undefined notions by other undefined notions, and nothing is gained. In fact, the modern geometer takes the notions of 'point' and 'straight line' as given, and does not attempt to define them. But Euclid's later definitions still make sense.

Notice that the definition of a right angle is entirely nonquantitative; no mention is made of 90° or $\pi/2$. For the Greeks, geometry was nonnumeric, being founded entirely on observation of patterns of shape; in particular, they regarded lengths and angles as geometric notions, not numeric ones.

Having defined, or at least attempted to define, the basic notions, Euclid's next step was to formulate five basic postulates, from which all geometric facts were supposed to follow by means of pure logical reasoning.

Postulate 1 [It is possible] to draw a straight line from any point to any point.

Postulate 2 [It is possible] to produce a finite straight line continuously in a straight line.

Postulate 3 [It is possible] to describe a circle with any center and [radius].

Postulate 4 All right angles are equal to one another.

Postulate 5 If a straight line falling on two straight lines makes the interior angles on the same side less than two right angles, the two straight lines, if produced indefinitely, meet on that side on which are the angles less than the two right angles. (See the figure on this page.)

In writing down these postulates, and then deducing other geometric facts from them, Euclid was not trying to establish some kind of logical 'game' with arbitrary rules, such as chess. For Euclid, and for generations of mathematicians after him, geometry was the study of regular shapes *as may be observed in the world*. The five postulates were supposed to be self-evident truths about the world, and in formulating them, Euclid was trying to capture some fundamental patterns of nature.

Euclid's postulates proved adequate for him to develop a great deal of geometry, in particular the forty-eight propositions in Book I of *Elements,* culminating in the Pythagorean theorem and its converse. Given the amount of geometry they produced, the five postulates are astonishingly few in number and, in all but one case, extremely simple in content.

Still, it is a bit of an overstatement to say, as I just did, that Euclid's axioms were 'adequate'. Along with many mathematicians after him, Euclid tacitly assumed a number of 'facts' that he did not formulate as postulates; for example:

- A straight line passing through the center of a circle must intersect the circle.

- A straight line that intersects one side of a triangle, but does not pass through any vertex of the triangle, must intersect one of the other sides.

- Given any three distinct points on the same line, one of them is between the other two.

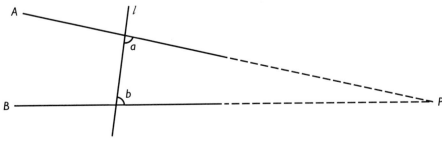

If line *l* meets lines *A* and *B* so that the sum of the angles *a* and *b* is less than two right angles, then *A* and *B* will meet, if extended far enough, on the *a,b* side.

The Golden Ratio

Mathematical patterns sometimes reflect visual patterns that the human eye finds particularly aesthetic. One famous example of such mathematical pattern is the *golden ratio*. This number is mentioned at the beginning of Book VI of Euclid's *Elements*.

According to the Greeks, the golden ratio is the ideal proportion for the sides of a rectangle that the eye finds the most pleasing. The rectangular face of the front of the Parthenon has sides whose ratio is in this proportion, and it may be observed elsewhere in Greek architecture. The golden ratio is also found in nature: The chambered shell of the Nautilus mollusk, pictured here, grows to form a logarithmic spiral, a mathematical curve that spirals out in a fashion dependent on the golden ratio.

The value of the golden ratio is $(1 + \sqrt{5})/2$, an irrational number approximately equal to 1.618. It is the number you get when you divide a line into two pieces so that the ratio of the whole line to the longer piece equals the ratio of the longer piece to the shorter.

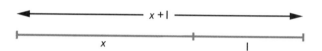

Expressing this algebraically, if the ratio concerned is $x{:}1$, as illustrated above, then x is a solution to the equation

$$\frac{x + 1}{x} = \frac{x}{1},$$

that is,

$$x^2 = x + 1.$$

The positive root to this equation is $x = (1 + \sqrt{5})/2$.

Since Euclid's aim in developing geometry in an axiomatic fashion was to avoid any reliance on a diagram during the course of a proof, it is perhaps surprising that he overlooked such basic assumptions as those just listed. On the other hand, his was the first serious attempt at axiomatization, and, when compared to the state of, say, physics or medicine in 350 B.C., Euclid's mathematics looks, and was, centuries ahead of its time.

Over two thousand years after Euclid, in the early part of the twentieth century, David Hilbert finally wrote down a list of twenty postulates that *is* adequate for the development of Euclidean geometry, in that all of the theorems in *Elements* can be proved from these postulates, using only pure logic.

The one postulate in Euclid's list that is clearly not 'simple' is the fifth one. Compared to the other four postulates, it is complicated indeed. Stripping

The golden ratio crops up in various parts of mathematics. One well-known example is in connection with the Fibonacci sequence. This is the sequence of numbers you get when you start with 1 and form the next number at each stage by adding together the two previous ones (except at step 2 where you only have one previous number). Thus, the sequence begins

$$1, 1, 2, 3, 5, 8, 13, 21, \ldots$$

This sequence captures a pattern that may be observed in many situations involving growth, from the growth of plants to the growth of a computer database. If $F(n)$ denotes the nth number of this sequence, then, as n becomes larger, the ratio

$$F(n + 1)/F(n)$$

of successive terms of the Fibonacci sequence gets closer and closer to the golden ratio.

The golden ratio has a particularly intriguing representation as a fraction that continues forever, namely:

$$1 + \cfrac{1}{1 + \cfrac{1}{1 + \cfrac{1}{1 + \cfrac{1}{\ddots}}}}$$

away the tangled verbiage, what it says is that if two straight lines in the plane are inclined toward one another, they will eventually meet. Another way to express this is that through a given point can be drawn exactly one straight line parallel to a given straight line. In fact, the fifth postulate looks more like a theorem than an axiom, and it seems that Euclid himself was reluctant to assume it, since he avoided any use of it in *Elements* until Proposition

I.29. Indeed, many subsequent generations of mathematicians attempted to deduce the fifth postulate from the other four, or to formulate a more basic assumption from which the fifth postulate could be deduced.

It was not that anyone doubted the veracity of the postulate. On the contrary, it seems perfectly obvious. It was its logical form that caused the problem: axioms were not supposed to be so specific or complicated. (It is not clear that such a view would prevail today; since the nineteenth century, mathematicians have learned to live with many more complicated axioms that likewise are supposed to capture 'obvious truths'.)

Still, obvious or not, no one was able to deduce the fifth postulate from the others, a failure that was taken to indicate a lack of understanding of the geometry of the world we live in. Nowadays, we recognize that there was indeed a failure of comprehension, but not of Euclidean geometry itself. Rather, the problem lay in the assumption that the geometry Euclid tried to axiomatize *was* the geometry of the world we live in—an assumption that the great philosopher Immanuel Kant, among others, took to be fundamental.

But that story will have to wait until a later section; in the meantime, we should take a look at some of the results Euclid obtained from his five postulates.

Euclid's Elements

Of the thirteen books that make up *Elements*, the first six are devoted to plane geometry in one form or another.

A number of the propositions in Book I concern 'ruler and compass' constructions. The task here is to determine what geometric figures can be constructed using just two tools, an unmarked straightedge, used only to draw straight lines, and a compass, which is used to draw arcs of circles but for no other purpose—in particular, the separation of the compass points is assumed to be lost when the instrument is taken from the page. Euclid's very first proposition describes just such a construction:

Proposition I.1   On a given straight line, to construct an equilateral triangle.

The method, illustrated on this page, seems simple enough. If the given line is AB, place the point of the compass at A and draw a quarter-circle above the line with radius AB, then place the point of the compass at B and draw a second quarter-circle with the same radius. Let C be the point where the two quarter-circles intersect. Then ABC is the desired triangle.

However, even here in the very first proposition, Euclid makes use of a tacit assumption that his axioms do not support: how do you know that the two quarter-circles intersect? Admittedly, the diagram suggests that they do, but diagrams are not always reliable; perhaps there is a 'hole' at C, much like the 'hole' in the rational line where $\sqrt{2}$ 'ought to be'. In any case, the whole point of writing down axioms in the first place was to avoid reliance on diagrams.

Other ruler-and-compass constructions include the bisection of an angle (Proposition I.9), the bisection of a line segment (Proposition I.10), and the construction of a perpendicular to a line at a point on the line (Proposition I.11).

It should be emphasized that, for all the attention given to them in *Elements*, Greek geometry was by no means restricted to ruler-and-compass constructions. Indeed, Greek mathematicians made use of whatever tools the problem seemed to demand. On the other hand, they did seem to regard ruler-and-compass constructions as a particularly elegant form of intellectual challenge: to the Greeks, a figure that could be constructed using only these two most primitive of tools was somehow more fundamental and pure, and a solution to a problem using just these tools was regarded as having particular aesthetic appeal. Euclid's postulates are clearly designed to try to capture what can be achieved using ruler and compass.

In addition to the results on constructions, Book I gives a number of criteria for establishing that two triangles are congruent (i.e. equal in all respects); for example, Proposition I.8 states that when two triangles have the three sides of one respectively equal to the three of the other, then the two triangles are congruent.

Book II deals with geometric algebra, establishing in a geometric fashion results that nowadays are generally handled algebraically, for example, the identity

$$(a + b)^2 = a^2 + 2ab + b^2.$$

Book III presents thirty-seven results about circles, including a proof that any angle inscribed in a semicircle is right. Euclid's elegant proof is given in the box on page 112.

Book IV includes constructions of regular polygons, polygons whose sides are all equal and whose angles are all equal, the simplest examples being the equilateral triangle and the square. Book V is devoted to an exposition of Eudoxus' theory of proportions, a geometrical theory designed to circumvent the difficulties raised by the Pythagoreans' discovery that $\sqrt{2}$ is not rational. This work was largely superseded by the nineteenth-century devel-

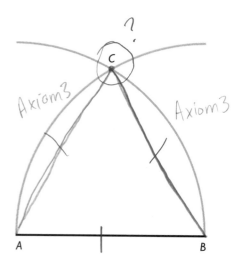

The construction of an equilateral triangle on a line.

A Gem from Euclid

Proposition I.47 of Euclid's *Elements* is the Pythagorean theorem. After he has completed the proof of that theorem, Euclid brings Book I to a close with the proof of the converse, namely:

Proposition I.48 If in a triangle the square on one of the sides be equal to the squares on the remaining two sides of the triangle, the angle contained by the remaining two sides of the triangle is right.

Proof Start with triangle ABC, in which it is assumed that

$$BC^2 = AB^2 + AC^2.$$

It is to be proved that $\angle BAC$ is right.

To do this, first draw AE perpendicular to AC at A, as shown in the diagram. This step is possible by Proposition I.11. Then construct the segment AD so that $AD = AB$. This is allowed by Proposition I.3.

The aim now is to show that triangles BAC and DAC are congruent; since $\angle DAC$ is right, it will follow at once that $\angle BAC$ is also a right angle.

The two triangles share a side, AC, and, by construction, $AD = AB$. By applying the Pythagorean theorem to the right-angled triangle DAC, you get

$$CD^2 = AD^2 + AC^2 = AB^2 + AC^2 = BC^2.$$

Hence $CD = BC$. But now the triangles BAC and DAC have equal sides, and hence, by Proposition I.8, they are congruent, as required. QED.

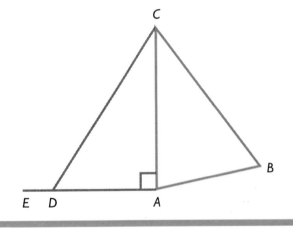

opment of the real-number system. The results in Book V are used in Book VI, in which Euclid presents a study of similar figures, two polygons being *similar* if their angles are respectively equal and the sides about equal angles are in proportion.

Book VI marks the end of Euclid's treatment of plane geometry. Books VII to IX are devoted to number theory, and Book X concerns measurement. Geometry becomes the focus once again in the final three books, this time the geometry of three-dimensional objects. Book XI contains thirty-nine propositions concerning the basic geometry of in-

tersecting planes. One major result (Proposition XI.21) says that at an apex of a polygonal-faced solid such as a pyramid, the sum of the plane angles converging at this point is less than four right angles.

In Book XII, the study commenced in Book XI is taken further with the aid of Eudoxus' method of exhaustion. Among the results proved is that the area of a circle is proportional to the square of the diameter (Proposition XII.2).

The final book of *Elements*, Book XIII, presents eighteen propositions on the regular solids. These are three-dimensional figures having planar faces,

A Theorem from Euclid's Elements

Proposition III.31   An angle inscribed in a semicircle is a right angle.

Proof   Draw a semicircle with center O and diameter BC, as in the diagram. Let A be any point on the semicircle. The theorem asserts that $\angle BAC$ is a right angle.

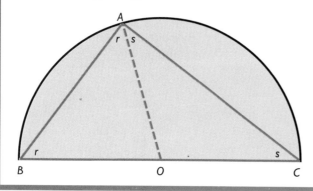

Draw the radius OA, and let $\angle BAO = r$, $\angle CAO = s$. Since AO and BO are radii of the semicircle, $\triangle ABO$ is isosceles; hence, as base angles of an isosceles triangle are equal, $\angle ABO = \angle BAO = r$. Similarly, $\triangle AOC$ is isosceles, and so $\angle ACO = \angle CAO = s$.

But the angles of a triangle sum to two right angles. Applying this fact to $\triangle ABC$,

$$r + s + (r + s) = \text{two right angles,}$$

which simplifies to give

$$2(r + s) = \text{two right angles.}$$

Thus $r + s$ is equal to one right angle. But $\angle BAC = r + s$, so the proof is complete. QED.

---

each face a regular polygon, all faces congruent, and all angles between pairs of adjacent faces the same. From early times, the Greeks had known of five such objects, illustrated on this page:

- the tetrahedron, having four faces, each one an equilateral triangle;

- the cube, having six square faces;

- the octohedron, having eight equilateral triangles as faces;

- the dodecahedron, having twelve regular pentagons for faces;

- the icosahedron, having twenty equilateral triangles as faces.

As a result of their being featured prominently in Plato's writings, the regular solids are sometimes re-

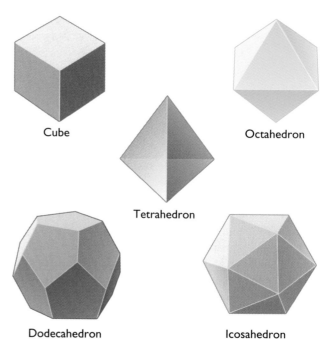

The five regular solids.

Plato's Atomic Theory

In *Timaeus*, written around 350 B.C., Plato put forward the theory that the four 'elements' that were believed to make up the world—fire, air, water, and earth—were all aggregates of tiny solids. Moreover, he argued, since the world could only have been made from perfect bodies, these elements must have the shape of regular solids. As the lightest and sharpest of the elements, fire must be a tetrahedron. As the most stable of the elements, earth must consist of cubes. Being the most mobile and fluid, water has to be an icosohedron, the regular solid most likely to roll easily. As to air, Plato observed that ". . . air is to water as water is to earth," and concluded, somewhat mysteriously, that air must be an octahedron. And finally, so as not to leave the one remaining regular solid out of the picture, he proposed that the dodecahedron represented the shape of the entire universe.

For all that this theory seems whimsical and fanciful to modern eyes, it was still taken seriously, even if not completely believed, in the sixteenth and seventeenth centuries, when Johannes Kepler began his quest for mathematical order in the world around him. The drawings shown are Kepler's own illustrations of Plato's atomic theory.

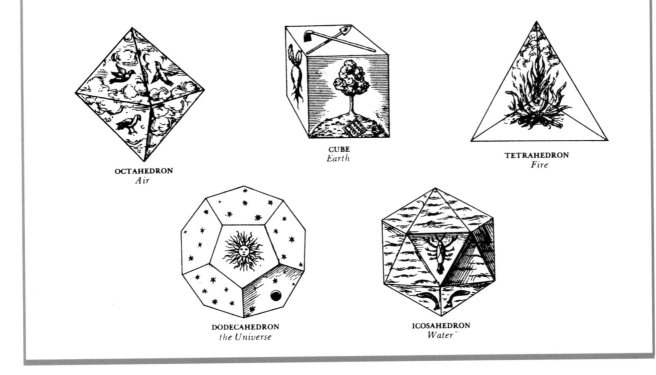

OCTAHEDRON
Air

CUBE
Earth

TETRAHEDRON
Fire

DODECAHEDRON
the Universe

ICOSAHEDRON
Water

ferred to as Platonic solids. The box on this page discusses the role played by these solids in early theories of the nature of the world.

The 465th, and last, proposition in *Elements* is Euclid's elegant proof that the five regular solids listed above are the only ones there are. All that is needed for the proof is Proposition XI.21, that the sum of the face angles at any apex is less than four right angles. The proof goes like this.

For a regular solid with triangular faces, the face angles are all 60°, and there must be at least three faces at each vertex. Exactly three triangles meeting

Kepler's Plantetary Theory

Somewhat more 'scientific' than Plato's atomic theory (described in the box on the previous page.) is Kepler's suggestion of the role played by the regular solids in the universe. There were six known planets at the time: Mercury, Venus, Earth, Mars, Jupiter, and Saturn. Influenced by Copernicus' theory that the planets move round the sun, Kepler tried to find numerical relations to explain why there were exactly six planets, and why they were at their particular distances from the sun. He decided eventually that the key was not numerical but geometric. There were precisely six planets, he reasoned, because the distance between each adjacent pair must be connected with a particular regular solid, of which there are just five.

After some experimentation, he found an arrangement of nested regular solids and spheres so that each of the six planets had an orbit on one of six spheres: the outer sphere (on which Saturn moves) contains an inscribed cube, and on that cube is inscribed in turn the sphere for the orbit of Jupiter. In that sphere is inscribed a tetrahedron; and Mars moves on that figure's inscribed sphere. The dodecahedron inscribed in the Mars-orbit sphere has the Earth-orbit sphere as its inscribed sphere, in which the inscribed icosahedron has the Venus-orbit sphere inscribed. Finally, the octa-

hedron inscribed in the Venus-orbit sphere has itself an inscribed sphere, on which the orbit of Mercury lies.

To illustrate his remarkable discovery, Kepler drew the painstakingly detailed picture reproduced on the facing page. Clearly, he was most pleased with what he had done. The only problem is that it is all nonsense!

First, the correspondence between the nested spheres and the planetary orbits is not really exact. Having himself been largely responsible for producing accurate data on the planetary orbits, Kepler was certainly aware of the discrepencies, and tried to adjust his model by taking the spheres to be of different thicknesses, though without giving any reason why the thicknesses should differ.

Second, as we now know, there are not six but at least nine planets. Uranus, Neptune, and Pluto were discovered subsequent to Kepler's time.

From the modern standpoint, it may seem hard to believe that two intellectual giants of the caliber of Plato and Kepler should have proposed such crackpot theories. What drove them to seek connections between the regular solids and the structure of the universe?

The answer is that they were driven by the same deep-seated belief that motivates today's scientist:

---

at a vertex gives an angle sum of 180°; exactly four triangles gives an angle sum of 240°; and exactly five triangles an angle sum of 300°. With six or more triangles meeting at a vertex, the angle sum would be equal to, or greater than, 360°, which is impossible by Proposition XII.21. So there are at most three regular solids with triangular faces.

For a regular solid with square faces, if three faces meet at a vertex, the angle sum at that vertex is 270°. But if four or more squares were to meet at a vertex, the angle sum would be 360° or more,

so that cannot occur. Hence there can be at most one regular solid having square faces.

The interior angle at a vertex of a pentagon is 108°, so the only possible regular solid with pentagonal faces will have three faces meeting at each vertex, with an angle sum of 324°.

The interior angle at a vertex of any regular polygon with six or more sides is at least 120°. Since 3 × 120° = 360°, Proposition XII.21 implies that there can be no regular solid whose faces have six or more sides.

that the pattern and order in the world can all be described, and to some extent explained, by mathematics. At the time, Euclid's geometry was the most well developed branch of mathematics, and the theory of the regular solids occupied a supreme position within geometry; a complete classification had been achieved, with all five regular solids having been been identified and extensively studied. Though ultimately not sustainable, Kepler's theory was, in its conception, extremely elegant, and very much in keeping with the view expressed by his contemporary Galileo: "The great book of nature can be read only by those who know the language in which it was written. And this language is mathematics." Indeed, it was Kepler's fundamental belief in mathematical order that led him to adjust his mathematical model in order to fit the observed data, pursuing the aesthetic elegance of the model even at the cost of a 'fudge' that he could not explain.

In the particulars, both Plato and Kepler were most certainly off target with their atomic theories. But in seeking to understand the patterns of nature through the abstract patterns of mathematics, they were working within a tradition that continues to this day to be highly productive.

These considerations imply that the five regular solids already listed are the only ones possible.

The regular solids continued to have an influence in scientific theorizing about the world, as illustrated in the box on this page.

One further piece of Greek geometry should be mentioned, though since it postdates Euclid's *Elements* by several generations, it cannot strictly be classified as Euclidean geometry. This is the work on the conic section contained in Apollonius' eight-volume treatise *Conics*. The conic sections are curves produced when a slice is cut through a cone, as illustrated in the figure on the next page. There are three such curves, the ellipse, the parabola, and the hyperbola. (The circle is a special case of an ellipse.) These curves had been extensively studied throughout the Greek era, but it was in *Conics* that this work was all brought together and systematized, in much the same way that *Elements* organized what was known in Euclid's time.

Along with *Elements* and the work of Archimedes, Apollonius' *Conics* was still in use as a highly

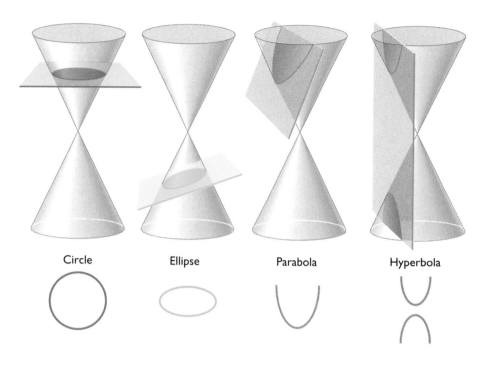

The conic sections. These four curves are obtained when a hollow double-cone is sliced by a plane.

Circle   Ellipse   Parabola   Hyperbola

regarded textbook in the seventeenth century, when Kepler made his profound observation that the planets travel round the sun in elliptical orbits. This discovery demonstrated that, far from being purely of aesthetic interest, one of the conic sections—the ellipse—was the very path through which the planets, and humanity along with them, travel through space.

In fact, the shape of planetary orbits is not the only instance where conic sections figure in the physics of motion; when a ball or other projectile is thrown into the air, it follows a parabolic path, a fact utilized by the mathematicians who prepare the artillery tables to ensure that wartime gunners can accurately aim their shells to hit the desired target.

An oft-repeated myth is that Archimedes made use of the properties of the parabola to defend the city of Syracuse against the Roman invaders in their war against Carthage. According to this story, the great Greek mathematician built huge parabolic mirrors that focused the sun's rays on the enemy ships, setting them ablaze. The mathematical property of the parabola involved here is that rays that fall on the parabola parallel to its axis are all re-

flected to a single point, known as the parabola's focus. This property is illustrated in the figure on this page. Given the difficulties that would be encountered in trying to aim such a device, this story is unlikely to be true. However, the same mathemat-

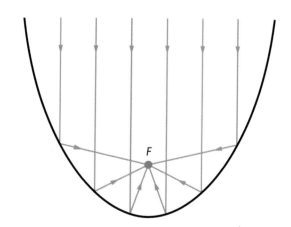

Lines approaching the parabola parallel to its axis are reflected to its focus (F). This geometric fact has found a number of applications, both in war and in peacetime.

French mathematician and philosopher René Descartes in 1637, when he introduced the idea of *co-ordinate geometry*. In coordinate geometry, figures can be described by means of algebraic equations. For example, the conic sections are precisely the curves that can be described by means of quadratic equations involving the two variables x and y. Descartes' contribution to geometry is the subject of the next section.

Cartesian Geometry

In 1637, Descartes published the book *Discours de la Méthode,* a highly original philosophical analysis of the scientific method. In an appendix, entitled *La Géométrie,* he presented the mathematical world with a revolutionary new way to do geometry: by algebra. Indeed, the revolution Descartes set in mo-

The Greek mathematician Archimedes, as represented in an early seventeenth-century painting by Domenico Fetti.

ical property is successfully used today in the design of automobile headlights, satellite dishes, and telescope reflectors.

At first glance, the three conic sections appear to be quite distinct kinds of curve, one being a closed loop, another a single arch, and the third consisting of two separate segments. It is only when you see how they are produced by taking slices through a double cone that it becomes clear that they are all part of a single family—that there is a single, unifying pattern. But notice that you have to go to a higher dimension to discover the pattern: the three curves all lie in the two-dimensional plane, but the unifying pattern is a three-dimensional one.

An alternative pattern linking the conic sections, an algebraic pattern, was discovered by the

René Descartes (1596–1650).

tion with this publication was to be a complete one in that, not only did his new approach enable mathematicians to make use of algebraic techniques in order to solve problems in geometry, it effectively gave them the option of regarding geometry *as a branch of algebra*.

Descartes' key idea was (taking the two-dimensional case for definiteness) to introduce a pair of coordinate axes, two real-number lines drawn at right angles, as shown in the figure on this page. The point of intersection of the two axes is called the origin (of coordinates). The two axes are most frequently labeled the x-axis and the y-axis; the origin is generally denoted by 0.

In terms of the coordinate axes, every point in the plane has a unique 'name' as a pair of real numbers, the x-coordinate of the point and the y-coordinate. The idea then is to represent geometric figures by algebraic expressions involving x and y; in particular, lines and curves are represented by algebraic equations in x and y.

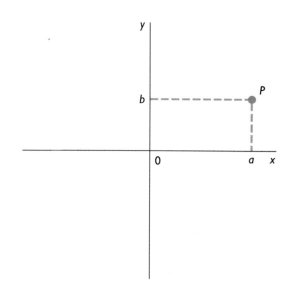

Relative to a given pair of Cartesian coordinate axes, every point in the plane has a unique name as a pair of real numbers. The point P shown has coordinates (a,b), where a,b are real numbers.

For example, a straight line with gradient m that crosses the y-axis at the point $y = c$ has the equation

$$y = mx + c.$$

The equation of a circle with center at the point (p,q) and radius r is

$$(x - p)^2 + (y - q)^2 = r^2.$$

Expanding the two bracketed terms in this equation and rearranging, the expression becomes

$$x^2 + y^2 - 2px - 2qy = k,$$

where $k = r^2 - p^2 - q^2$. In general, any equation of this last form for which $k + p^2 + q^2$ is positive will represent a circle with center (p,q) and radius $\sqrt{k + p^2 + q^2}$. In particular, the equation of a circle of radius r whose center is at the origin is

$$x^2 + y^2 = r^2.$$

An ellipse centered about the origin has an equation of the form

$$\frac{x^2}{a^2} + \frac{y^2}{b^2} = 1.$$

A parabola has an equation of the form

$$y = ax^2 + bx + c.$$

Finally, a hyperbola centered about the origin has an equation of the form

$$\frac{x^2}{a^2} - \frac{y^2}{b^2} = 1$$

or (a special case) of the form

$$xy = k.$$

Some of these curves are illustrated on the facing page.

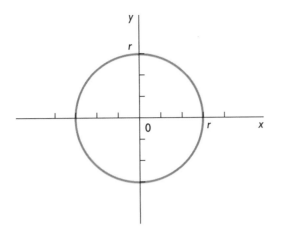

The circle $x^2 + y^2 = r^2$.

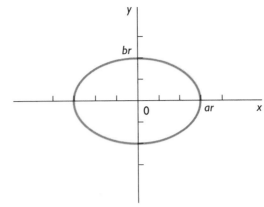

The ellipse $\dfrac{x^2}{a^2} + \dfrac{y^2}{b^2} = r^2$.

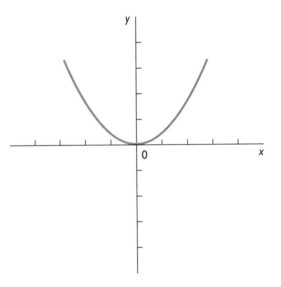

The parabola $y = x^2$.

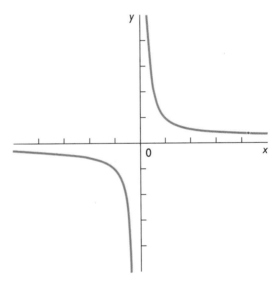

The hyperbola $xy = r^2$.

When algebraic equations are used to represent lines and curves, geometric arguments such as the ones put forth by the ancient Greeks may be replaced by algebraic operations, such as the solution of equations. For instance, determining the point at which two curves intersect corresponds to finding a common solution to the two equations. In order to find the points P, Q where the line $y = 2x$ meets the circle $x^2 + y^2 = 1$ (as in the figure on the next page), you solve these two equations simultaneously.

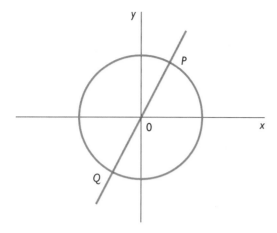

To find the points of intersection P and Q, one solves simultaneous equations for the circle and the line.

Substituting $y = 2x$ in the second equation gives $x^2 + 4x^2 = 1$, which solves to give $x = \pm 1/\sqrt{5}$. Using the equation $y = 2x$ to obtain the corresponding y-coordinates of the points of intersection, you get $P = (1/\sqrt{5}, 2/\sqrt{5})$, $Q = (-1/\sqrt{5}, -2/\sqrt{5})$.

Again, the perpendicularity of two straight lines corresponds to a simple algebraic condition on their equations. The lines $y = mx + c$ and $y = nx + d$ are perpendicular if, and only if, $mn = -1$.

By incorporating methods of the differential calculus, questions to do with lines tangent to curves may also be dealt with algebraically. For instance, the tangent line to the curve $y = f(x)$ at the point $x = a$ has the equation

$$y = f'(a)x + [f(a) - f'(a)a].$$

The use of algebraic and calculus-based analytic techniques in geometry provides a degree of precision, and a potential for greater abstraction, that takes the study of 'shape' into realms that would otherwise remain forever inaccessible. One early illustration of the power of these techniques was the resolution, toward the end of the nineteenth century, of three geometric problems that had defied solution since the time of the Greeks.

In addition to enabling generations of mathematicians to tackle geometric questions using the techniques of algebra, Cartesian geometry underpins the present-day technology of computer graphics. The illustration shows a computer animation scene from the motion picture *The Lawnmower Man*. The creatures that appear on the screen exist only as collections of mathematical equations in the computer memory.

Three Classic Problems

Computing the area of a square or rectangle whose dimensions are known is an easy matter involving nothing more complicated than multiplication. On the other hand, to compute the area of a figure with curved edges, such as a circle or an ellipse, seems far more difficult. The Greeks used the method of exhaustion, present-day mathematicians use the integral calculus. Both techniques are considerably more complicated than multiplication.

Another possible way is to find a square whose area is the same as the curved figure, and then compute the area of that square in the normal fashion. Can such a square be found, and if so, how? This is the problem of *quadrature* of a given figure, a problem the Greeks spent considerable time trying to solve. The simplest case, at least to state, is the problem of quadrature of the circle: given a circle, find a square having the same area.

Not surprising, the Greeks wondered if there were a construction using only ruler and compass, the 'pure' tools favored by Euclid in *Elements*. They were unable to produce an answer. Nor did countless succeeding generations of geometers fare any better, either the professional mathematicians or any of the many amateur would-be 'circle-squarers' who tried their hands at the problem. It should be noted that the problem asks for an *exact* solution; there are a number of ruler-and-compass methods for obtaining approximate solutions.

In 1882, the German mathematician Ferdinand Lindemann finally brought the quest to an end, by proving conclusively that ruler-and-compass quadrature of the circle is impossible. His proof was purely algebraic, by way of Cartesian coordinates, and goes like this.

First of all, operations performed with ruler and compass have corresponding algebraic descriptions. Fairly elementary considerations show that any length that can be constructed using ruler and compass can be computed, starting from integers, using only a sequence of additions, multiplications, subtractions, divisions, and the extraction of square roots. It follows that any length that can be constructed using the classic Greek tools must be *algebraic*; that is, it can be obtained as the solution to a polynomial equation

$$a_nx^n + a_{n-1}x^{n-1} + \cdots + a_2x^2 + a_1x + a_0 = 0,$$

where the coefficients $a_n, \ldots, a_0$ are all integers.

Real numbers that are not algebraic are called *transcendental*. Transcendental numbers are all irrational. (The converse is not true; for instance, $\sqrt{2}$ is irrational, but is certainly not transcendental, being a solution to the equation

$$x^2 - 2 = 0.)$$

What Lindemann did was to prove that the number π is transcendental. His proof used methods of the calculus.

That π is transcendental at once implies that quadrature of the circle is impossible. For consider the unit circle, the circle with radius $r = 1$. The area of the unit circle is $\pi \times r^2 = \pi \times 1 = \pi$. So, if you could construct a square having the same area as the unit circle, the square would have area π, and its edges would have length $\sqrt{\pi}$. In this way, therefore, you would be able to construct a length of $\sqrt{\pi}$ using ruler and compass. This would imply that $\sqrt{\pi}$ was algebraic, and it would follow at once that π was algebraic, contrary to Lindemann's result.

Another Greek problem that was resolved with the aid of Cartesian techniques was that of *duplicating the cube*: given any cube, find the side of another cube having exactly twice the volume. Once again, the problem was to find a construction using ruler and compass only, and once again, the problem can be solved if this restriction is relaxed. In particular, duplication of the cube is easily accomplished by means of a so-called neusis construction, which is performed using a compass and a *marked* ruler that may be slid along some configuration of lines until a certain condition is met. Another solution uses conic sections, and there is a three-dimensional construction involving a cylinder, a cone, and a torus.

If you can duplicate the unit cube, which has volume 1, then the duplicate has volume 2, and hence its edges must have length $\sqrt[3]{2}$. In algebraic terms, therefore, duplication of the unit cube corresponds to solving the cubic equation $x^3 = 2$. An argument somewhat simpler than Lindemann's shows that this equation cannot be solved by means of a sequence of additions, multiplications, subtractions, divisions, and the extraction of square roots, the operations that correspond to ruler-and-compass constructions. Hence the unit cube cannot be duplicated using only ruler and compass.

A third Greek problem that achieved some fame as a result of defying solution for many years was that of trisecting an angle. The problem is to find a method, using ruler and compass only, that, for any given angle, will produce an angle exactly one-third the size.

The problem can be solved immediately for some angles; for example, trisecting a right angle is easy: you simply construct a $30°$ angle. But the problem asks for a method that works in all cases. It is possible to solve the general problem if the restriction to ruler and compass is relaxed; in particular, trisection of any angle is easy by means of a neusis construction.

Again, in algebraic terms, trisection of an arbitrary angle amounts to the solution of a cubic equation, which is not possible using only the basic arithmetic operations together with the extraction of square roots. Thus, trisection cannot be achieved using ruler and compass alone.

It should be re-emphasized that the above three problems were not, in themselves, 'major' mathematical problems. Restriction to ruler-and-compass constructions was largely a Greek intellectual game. The Greeks could solve all three problems when this restriction was removed. It was the fact that the problems as stated defied solution for so long that led to their fame.

It is interesting that, in each case, the solution came after the problem was translated from a purely geometric task to an algebraic one, a transformation that enabled other techniques to be brought to bear.

As originally formulated, the three problems concerned patterns (i.e. sequences) of geometric construction using certain tools; the solutions depended on reformulating the problems in terms of equivalent, algebraic patterns.

Non-Euclidean Geometries

As mentioned on page 109, from the first formulation of Euclid's postulates for geometry, the fifth, or 'parallel postulate', was regarded as problematic. Its truth was never questioned—everyone agreed it was 'obvious'. But mathematicians felt that it was not sufficiently fundamental to be taken as an axiom; rather it ought to be proved as a theorem.

The 'obviousness' of the axiom seemed to be emphasized by the various alternative formulations that were obtained. In addition to Euclid's original statement of the fifth postulate—one of the least obvious formulations as it happens—each of the following can be shown to be completely equivalent to the fifth postulate:

Playfair's postulate: Given a straight line and a point not on the line, exactly one straight line may be drawn through the point, parallel to the given line. (The formal definition of 'parallel' is that two lines are parallel if, however far they be extended, they do not meet.)

Proclus' axiom: If a line intersects one of two parallel lines, it must intersect the other also.

Equidistance postulate: Parallel lines are everywhere equidistant.

Triangle postulate: The sum of the angles of a triangle is two right angles.

Triangle area property: The area of a triangle can be as large as we please.

Three points property: Three points either lie on a straight line or lie on a circle.

Most people find at least one or more of these statements 'obvious', generally the first three in the list, and maybe also the triangle area property. But why do they? Take the Playfair postulate as an example. How do you know it is true? How could you test it?

Suppose you drew a line on a sheet of paper and marked a point not on the line. You are now faced with the task of showing that there is one and only one parallel to the given line that passes through the chosen point. But there are obvious difficulties here. For one, no matter how fine the point of your pencil, the lines you draw still have a definite thickness, and so how do you know where the actual *lines* are? Second, in order to check that your second line is in fact parallel to the first, you would have to extend both lines indefinitely, which is not possible. Certainly, you can draw many lines through the given point that do not meet the given line *on the paper*.

Thus, Playfair's postulate is not really suitable for experimental verification. How about the triangle postulate? Certainly, verifying this postulate does not require extending lines indefinitely; it can all be done 'on the page'. Admittedly, it is likely that no one has any strong intuition concerning the angle sum of a triangle being 180°, the way we do about the existence of unique parallels, but since the two statements are entirely equivalent, the absence of any supporting intuition does not affect the validity of the triangle approach.

Suppose, then, that you could draw triangles and measure their three angles with sufficient accuracy that you could calculate the sum of the angles to within 0.001 of a degree, a perfectly reasonable assumption in this day and age. You draw a triangle and measure the angles. If these add up to 180°, then all you can conclude for certain is that the angle sum is between 179.999° and 180.001°, which is inconclusive.

On the other hand, it is, in principle, possible to demonstrate the falsity of the fifth postulate in this way. If you found a triangle whose angle sum worked out to be 179.9°, then you would know for

certain that the angle sum was between 179.899° and 179.901°, and so the answer could not possibly be 180°.

According to mathematical folklore, Gauss himself made an attempt to decide the fifth postulate experimentally, in the early part of the nineteenth century. To avoid the problem of drawing infinitely thin straight lines, he took light to be the straight lines, and to minimize the effect of errors in measurement, he worked with a very large triangle, whose apexes were located on three mountaintops. Setting up a fire on one of them, and using mirrors to reflect the light, he created a huge triangle of light rays. The angle sum worked out to be 180° plus or minus the experimental error. With the greatest optimism in the world, that experimental error could be no less than 30 angular seconds. So the experiment could have proved nothing, except that on a scale of several miles, the angle sum of a triangle is *fairly close* to 180°.

In fact, there seems to be no sound basis in our everyday experience to support the fifth postulate. And yet, in a form such as Playfair's postulate, we do believe it, indeed we find it obvious, almost every one of us. The abstract concept of a 'straight line', having length but no width, seems perfectly sound, and we can in fact visualize such entities; the idea of two such lines being indefinitely extended, and remaining everywhere equidistant and not meeting, seems meaningful; and we have this deep-rooted sense that parallels exist and are unique.

As I tried to indicate at the very beginning of this chapter, these fundamental geometric notions, and the intuitions that accompany them, are not part of the physical world we live in; they are part of ourselves, of the way we are constructed as cognitive entities. Euclidean geometry may or may not be the way the world is 'made up', whatever that may mean; but it does appear to capture the way human beings *perceive* the world.

So where does all of this leave the geometer? On what does she base her subject, a subject that deals with 'points', 'lines', 'curves', and the like, all of which are abstract idealizations, molded by our perceptions? The answer is that, when it comes to establishing the theorems that represent mathematical truths, the axioms are, quite literally, *all* there is. Practical experience and physical measurement cannot give us the certainty of mathematical knowledge. In constructing a proof in geometry, we might well rely upon mental pictures of straight lines, of circles, and so forth, in order to guide our reasoning process, but our proof must rest entirely upon what the axioms tell us about those entities.

Despite Euclid's attempt to axiomatize geometry, it took mathematicians two thousand years to come fully to terms with the significance of this last remark, and to abandon the intuition that told them that Euclidean geometry was self-evidently the geometry of the universe we live in.

The first significant step toward this realization was made by an Italian mathematician called Girolamo Saccheri. In 1733, he published a two-volume work titled *Euclid freed of every flaw*. In this book, he tried to establish the fifth postulate by showing that its negation resulted in a contradiction.

Given a line and a point not on the line, there are three possibilities for the number of parallel lines through that point:

(i) there is exactly one parallel;

(ii) there are no parallels;

(iii) there is more than one parallel.

These are illustrated in the figure on this page.

Possibility (i) is Euclid's fifth postulate. Saccheri set out to demonstrate that each of the other two possibilities leads to a contradiction. Assuming that Euclid's second postulate requires straight lines to be infinitely long, he found that possibility (ii) leads to a contradiction. He was less successful in eliminating possibility (iii). He obtained a number of consequences of statement (iii) that were counterintuitive, but was unable to formally derive a contradiction.

A hundred years later, four different mathematicians, all working independently, tried the same

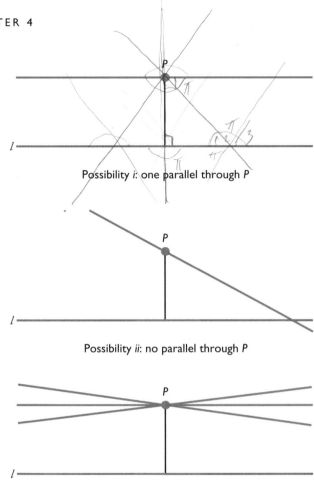

Possibility *i*: one parallel through *P*

Possibility *ii*: no parallel through *P*

Possibility *iii*: many parallels through *P*

The parallel postulate. Given a line *l* and a point *P* not on *l*, there are three possibilities for the existence of lines through *P* that are parallel to *l*: exactly one parallel, no parallels, or more than one parallel. The parallel postulate says that the first possibility occurs.

approach. But they made a crucial additional step that their predecessor had not: they managed to break free of the belief that there is only one geometry, the one in which the fifth postulate holds.

The first was Gauss. Working with the equivalent formulation of possibility (iii), that the angle sum of any triangle is less than two right angles, he came to realize that (iii) probably does not lead to an inconsistency, but rather to a strange, alternative geometry—a *non-Euclidean* geometry. Exactly when Gauss did this work is not known, since he did not publish it. The first reference we have is in a private letter written to a colleague, Franz Taurinus, in 1824, where he says,

> The assumption that the sum of the three angles is less than 180° leads to a curious geometry, quite different from ours, but thoroughly consistent, which I have developed to my entire satisfaction.

In another letter, written in 1829, he makes it clear that his reason for withholding his findings from publication is his fear that it would harm his considerable reputation if he were to go on record as saying that Euclidean geometry is not the only one possible.

Such are the pressures on the most successful. No such constraint hampered János Bolyai, a young Hungarian artillery officer, whose father, a friend of Gauss, had himself worked on the parallel axiom. Though the senior Bolyai advised his son not to waste his time on the problem, János took no notice, and was able to make the bold step his father had not. As Gauss had done, he recognized that assumption (iii) leads not to an inconsistency but to a completely new geometry. It was not until János' work was published as an appendix to a book of his father's in 1832 that Gauss informed the two men of his own earlier observations on the subject.

Sadly for Bolyai, not only had Gauss beaten him to the idea, and not only was he never widely recognized for his work during his lifetime, but it turned out that he was not even the first to publish on the new non-Euclidean geometry. Three years previously, in 1829, Nikolay Lobachevsky, a lecturer at Kazan University in Russia, had published much the same results under the title (translated from Russian) *Imaginary Geometry*.

So now there were two geometries, Euclidean geometry, with unique parallels, and another geometry with multiple parallels, known nowadays as hyperbolic geometry. It was not long before a third was added. In 1854, Bernhard Riemann re-examined the contradiction that Saccheri had derived from statement (ii), the statement that there are no parallels to a given line through a given point. He observed that this contradiction could be avoided. The crucial mistake Saccheri had made was to assume that Euclid's second postulate (a finite straight line can be produced continuously) implies the infinitude of the line. This assumption is simply not valid.

Riemann suggested that statement (ii) may be consistent with Euclid's first four postulates, and, if so, then when all five postulates are taken together as axioms, the result is yet another geometry, Riemannian geometry, in which the angles of a triangle always add up to more than two right angles.

None of the actors in this drama suggested that Euclidean geometry was not the right one for the

Nikolay Ivanovich Lobachevsky (1792–1856).

universe we live in. They simply advocated the view that the first four of Euclid's postulates do not decide between the three possibilities (i)–(iii), each one of which leads to a consistent geometry. Moreover, none of them provided a *proof* that (ii) or (iii) were consistent with Euclid's other postulates; rather they formed the *opinion* that this was so, based on their work with the postulate concerned. Proof that the two non-Euclidean geometries are consistent (more precisely, that they are as consistent as Euclidean geometry itself) came in 1868, when Eugenio Beltrami showed how to interpret each of the hyperbolic and Riemannian geometries within Euclidean geometry.

It is at this point that we must really come to grips with what it means to abandon one's intuition and work in a purely axiomatic fashion.

The basic, undefined objects in Euclidean geometry are points and straight lines. (Circles may be defined; a circle is the collection of all points equidistant from a given point.) We may well have various mental pictures and intuitions about these objects, but unless these pictures or intuitions are captured by the axioms, they are logically irrelevant as far as the geometry is concerned.

For example, consider the geometry that holds on the surface of the earth, which we shall assume to be a perfect sphere; call this geometry spherical geometry. Though the earth is a three-dimensional object, its surface is two-dimensional, so its geometry will be a two-dimensional geometry. What are the 'straight lines' in this geometry? The most reasonable answer is that the straight lines are the surface geodesics; that is, the straight line from a point *A* to a point *B* is the path of shortest distance from *A* to *B* on the surface. For a sphere, the straight line from *A* to *B* is the great circle route from *A* to *B*, illustrated in the first figure on this page. Viewed from above the earth, such a path does not appear to be a straight line, since it follows the curve of the earth's surface. But as far as the geometry *of the surface itself* is concerned, it appears to have all the properties of a straight line. A railway track that ran in a straight line from New York to San Francisco

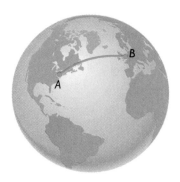

Straight lines in spherical geometry are arcs of great circles.

would follow such a path, and airplanes follow such paths all the time.

Spherical geometry satisfies the first four of Euclid's postulates, though in the case of the second postulate, notice that the continuous extendibility of a straight line does not imply infinite. Rather, when a straight line segment is extended, it eventually wraps around the globe and meets up with itself. The fifth postulate is not satisfied. Indeed, in this geometry there are no parallels; any two straight lines meet at two antipodal points, as shown in the second figure on this page.

Thus, the axioms of Riemannian geometry are true of the geometry of the surface of a sphere. At least, they are true given the way Euclid stated his

There are no parallels in spherical geometry; any two straight lines meet in a pair of antipodal points.

postulates. However, it is generally assumed that Euclid intended his first postulate to imply that two points determine *exactly one* straight line, and the same should be true of Riemannian geometry. Spherical geometry does not have this property; any pair of antipodal points lie on infinitely many straight lines, namely every great circle through those two points. To obtain a world in which this strengthening of the first postulate is valid, you declare that antipodal points are one and the same point. Thus, the North Pole is declared to be exactly the same point as the South Pole, any point on the equator is identified with the diametrically opposite point on the equator, and so on. Of course, this process creates a geometric 'monster' that defies visualization. But for all that you cannot picture this world, it provides a mathematically sound interpretation of Riemannian geometry: all the axioms, and hence all the theorems, of Riemannian geometry are true in this world, where the 'points' of the geometry are the points on the surface of the 'what was a sphere' and the 'straight lines' of the geometry are what you obtain from the great circles on the sphere when the antipodal points are declared identical.

Though the Riemannian world just described appears to defy visualization, working with its geometry is not so bad in practice. A difficulty only arises when you try to think of the entire world at once. The strange identification of antipodal points is only necessary to avoid problems that arise with 'infinitely long' lines, that is, lines that stretch half way round the globe. On a smaller scale, there is no such problem, and the geometry is the same as spherical geometry, a geometry that globe-trotting inhabitants of the planet earth are very familiar with. In particular, our everyday knowledge of global travel enables us to observe, as an empirical fact, the theorem of Riemannian geometry that the angles of a triangle add up to more than two right angles. In fact, the larger the triangle, the greater the angle sum. For an extreme case, imagine a triangle that has one apex at the North Pole and the other two apexes on the equator, one on the Green-

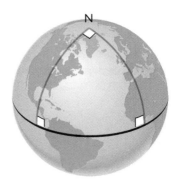

In spherical geometry, the angle sum of a triangle is more than 180°.

wich meridian and the other at 90° west, as shown in the figure on this page. Each angle of this triangle is 90°, so the angle sum is 270°. (Notice that in the sense of spherical geometry, this is a triangle; each side is a segment of a great circle, and hence the figure is bounded by three straight lines in the geometry.)

The smaller the triangle, the smaller is the angle sum. Triangles that occupy a very small area of the globe have angle sums very close to 180°. Indeed, for human beings living on the face of the earth, a triangle marked out on the surface will appear to have an angle sum of exactly 180°; the reason is that the effect of the earth's curvature is effectively zero on such a small scale. But, mathematically speaking, in spherical geometry, and likewise in Riemannian geometry, the angle sum of a triangle is never equal to 180°, it is always greater.

Turning now to hyperbolic geometry, the general idea is the same—to take the geometry of an appropriate surface, with the geodesics as the straight lines. The question is, what surface will produce hyperbolic geometry? The answer turns out to involve a pattern that is familiar to every parent.

Watch a child walking along, pulling a toy attached to a string. If the child makes an abrupt left turn, the toy will trail behind, not making a sharp corner, but curving around until it is almost behind the child once again. This curve is called a tractrix.

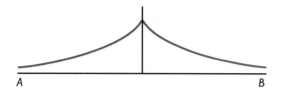

A double tractrix.

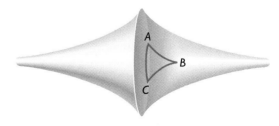

A pseudosphere.

Now take two opposite copies of a tractrix, as shown in the figure on the left, and rotate this double curve about the line *AB*. The resulting surface, shown on the right, is called a pseudosphere. It extends out to infinity in both directions.

The geodesic geometry on the pseudosphere is hyperbolic geometry. In particular, Euclid's first four postulates are true for this geometry, as are the other Euclidean axioms that Hilbert wrote down. But the fifth postulate is false: on the pseudosphere, given any line and any point not on that line, there are infinitely many lines through that point that are parallel to the given line, lines that will never meet, no matter how far extended.

The angle sum of any triangle drawn on a pseudosphere is less than 180°, as the figure at the bottom of this page illustrates. For very small triangles, where the curvature of the pseudosphere does not make much difference, the angle sum is close to 180°. But if you take a triangle and then start to enlarge it, the angle sum becomes less and less.

So, with three equally consistent geometries available, which one is the 'correct' one, the one chosen by nature? What is the geometry of our universe? It is not clear that this question has a single, definitive answer. The universe is the way it is; geometry is a mathematical creation of the human mind that reflects certain aspects of the way we encounter the environment. Why should the universe *have* a geometry at all?

Let's try to rephrase the question. Given that mathematics provides human beings with a very powerful means to describe and understand aspects

of the universe, which of the three geometries is best suited to that task? Which geometry most closely corresponds to the observable data?

On a small, human scale, a scale that covers part or all of the earth's surface, Newtonian physics provides a theoretical framework in complete accord with the observable (and measurable) evidence, and any one of the three geometries will do. Newton's physics depends upon Euclidean geometry. Since Euclidean geometry also seems to accord with our own intuitions about the way we perceive the world, we may as well take that to be 'the geometry of the physical world'.

On the other hand, at a larger scale, from the solar system up to galaxies and beyond, Einstein's relativistic physics provides a closer fit with the observable data than does Newton's framework. At this scale, non-Euclidean geometry appears to be more appropriate. According to relativity theory, space-time is curved, the curvature manifesting itself in what we refer to as the force of gravity. The curvature of space-time is observed in the behavior of

On a pseudosphere, the angle sum of a triangle is less than 180°.

Circle limit III, by the Dutch artist M. C. Escher. The interior of the circle is a world in which the geometry is hyperbolic. Within this world, all the creatures shown are the same size. The apparent decrease in size as the circumference is approached is due to the way this hyperbolic world is embedded in the Euclidean geometry of the page on which it appears. Escher was a prolific artist, who throughout his career was fascinated by mathematical ideas. His knowledge of geometry was considerable, and he wove geometric themes into many of his etchings and woodcuts. Escher based this particular woodcut on a circular model of hyperbolic geometry invented by Henri Poincaré.

light rays, the 'straight lines' of the physical universe. When light rays from a distant star pass close to a large mass, such as the sun, their path is 'bent round', just as the geodesics on a globe curve round the surface.

Which non-Euclidean geometry one takes depends on one's choice of a theory of the universe. If you assume that the present expansion of the universe will come to a halt, to be replaced by a contraction, Riemannian geometry is the most appropriate. If, on the other hand, you take the view that the universe will expand forever, then hyperbolic geometry is more suitable.

It is particularly fascinating that Einstein's relativity theory, and the astronomical observations that demonstrated its superiority over the Newtonian theory, came more than a half century after the development of non-Euclidean geometry. Here we have an example of how mathematics can move ahead of our understanding of the world. The initial abstraction of geometric pattern from obervations of the world around them led the Greeks to the development of a rich mathematical theory, Euclidean geometry. During the nineteenth century, purely mathematical questions about that theory, questions concerning axiomatics and proof, then led to the discovery of other geometric theories. Though these alternative geometries were initially developed purely as abstract, axiomatic theories, seemingly having no applications in the real world, it turned out that they were in fact more appropriate for the study of the universe on a large scale than was Euclidean geometry.

Projective Geometry

For the surveyor mapping out the local terrain or the man setting out to build a house, Euclidean geometry captures the relevant patterns of shape. For the sailor or the airline pilot circumnavigating the globe, spherical geometry is the appropriate framework. For the astronomer, the geometric patterns that arise may be those of Riemannian geometry. It is all a question of what you want to do and how you want to do it.

Renaissance artists, among them Leonardo Da Vinci and Albrecht Dürer, wanted to accurately portray depth on a two-dimensional canvas. Prior to the Renaissance period, it presumably never occurred to artists that there might be a means of rendering depth in their paintings.

The key idea exploited by these two artists is to consider the surface of a painting to be a window through which the artist views the object to be

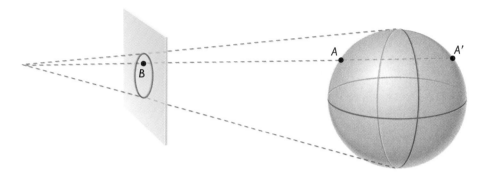

Projection. The lines of vision from the sphere to the eye form a cross section on the plane. Points A and A′ both *project* to the point B. The sphere *projects* to a circular region of the plane.

painted. The lines of vision from the object that converge to the eye pass through that window, and the points where the lines intersect the window surface form the *projection* of the object onto that surface as illustrated in the figure at the top of this page. The painting captures that projection, as illustrated by Dürer in the drawing shown below. For the artist, therefore, the relevant patterns are those of perspective and of plane projections; the geometry that

arises from these considerations is called projective geometry.

Though the fundamental ideas of perspective were discovered in the fifteenth century, and gradually came to pervade the world of painting, it was not until the eighteenth century that projective geometry was studied as a mathematical discipline. In 1813, while a prisoner of war in Russia, Jean-Victor Poncelet, a graduate of the École Polytech-

Albrecht Dürer's woodcut *Institutiones Geometricae* illustrates how a perspective drawing is produced by means of a projection. The glass plate held by the man on the left bears a perspective drawing of the object on the table. To create this drawing, the plate is placed in the frame held by the artist, on the right. Where the light rays (straight lines) from the object to the artist's eye intersect the plate, they determine an image of the object, known as its projection onto the plate.

The Annunciation, The Master of the Barberini, 1450. This painting from the Renaissance period exhibits single-point perspective, where all the perspective lines converge to a single 'point at infinity'.

nique in Paris, wrote the first book on the subject, *Traité des propriétés projectives des figures.* During the early nineteenth century, projective geometry grew into a major area of mathematical research.

If Euclidean geometry corresponds to our mental conception of the world around us, projective geometry captures some of the patterns that enable us to see the world the way we do, for our entire visual input consists of flat, two-dimensional images on our retinas. When an artist draws a painting that is correctly in perspective, we are able to interpret it as a three-dimensional scene, as, for example, the Renaissance painting above. Indeed, this visual capacity can be exploited to create a visual illusion of various 'impossible figures', such as the one shown on the next page. The mathematician is interested in the geometrical patterns in the painting that enable it to create the illusion of a three-dimensional scene.

The basic idea of projective geometry is to study those figures, and those properties of figures, that are left unchanged by a projection. For example,

An impossible figure from M. C. Escher.

points project to points and straight lines project to straight lines, so points and lines figure in projective geometry.

Actually, you have to be a little bit careful, since there are two kinds of projection. First, there is pro-jection from a single point, also known as *central projection*, which is illustrated in the figure below. This is the kind of projection used by the painter, where the projection point is the painter's eye. Then there is *parallel projection,* sometimes referred to as projection from infinity. This form of projection is illustrated in the figure on the facing page. This would correspond to the hypothetical case, where the painter's eye was infinitely far away from the scene and the canvas. In both illustrations, the point P is projected to the point P' and the straight line l is projected to the straight line l'. (In the axiomatic study of projective geometry, the distinction between these two kinds of projection disappears, since in the formal theory there are no parallel lines—any two lines meet, and what were once thought of as parallel lines become lines that meet at a 'point at infinity'.)

Clearly, projection distorts lengths and angles, in a way that depends on the relative positions of the objects depicted. Thus projective geometry cannot involve axioms or theorems about length, angle, or congruence. In particular, though the notion

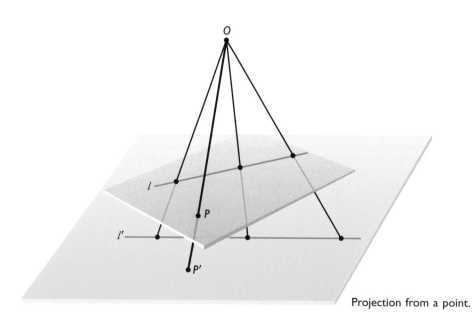

Projection from a point.

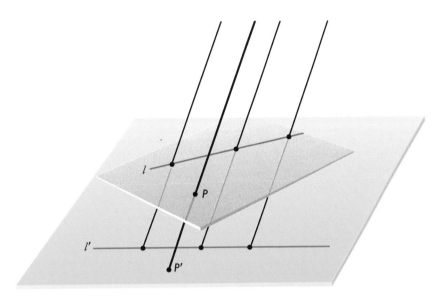

Parallel projection.

of a triangle makes sense in projective geometry, the notions of isosceles and equilateral triangles do not.

However, the projection of any curve is another curve, and this raises the interesting question of what classifications of curves are meaningful in projective geometry. For example, the notion of a circle is obviously not one of projective geometry, since the projection of a circle need not be a circle: circles often project into ellipses. However, the notion of a conic section is meaningful in projective geometry. (In Euclidean geometry, a conic section may be defined as a projection of a circle onto a plane. Different kinds of projection give rise to the different kinds of conic section.)

For this account, I shall concentrate on points and straight lines. What kinds of properties of points and lines make sense in projective geometry? What are the relevant patterns?

Obviously, the incidence of a point and a line is not changed by projection, so you can talk about points lying on certain lines and lines passing through certain points. An immediate consequence of this is that, in projective geometry, it makes sense to talk about three points lying on a straight line or of three lines meeting at a single point.

Is this all that can be said? A skeptical reader might begin to wonder if we have allowed ourselves anything like enough machinery to prove any interesting and nontrivial theorems. Certainly, we have thrown away most of the concepts that dominate Euclidean geometry. On the other hand, the geometry that remains captures the patterns that our eyes are able to interpret as perspective, so projective geometry cannot be without content. The issue is, does this content manifest itself in the form of interesting geometric theorems?

An immediate answer to this question is provided by a striking result proved in the early seventeenth century by the French mathematician Gérard Desargues:

Desargues' theorem   If in a plane two triangles ABC and $A'B'C'$ are situated so that the straight lines joining corresponding vertices all meet at a point O, then the corresponding sides, if extended, will intersect in three points that lie on the same line.

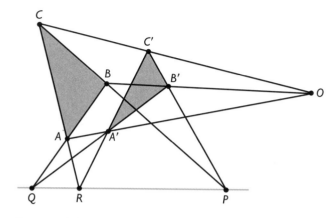

Desargues' theorem in the plane.

The figure on the right illustrates the theorem. If you doubt the result, draw a few further figures until you are convinced. It is clearly a result of projective geometry, since it speaks only of points, straight lines, triangles, lines meeting at a point, and points lying on a line, all concepts of projective geometry.

Cross-ratio

Given three points A, B, C on a straight line, a projection will send A, B, C to three points A', B', C' that are also on a straight line, but will, in general, change the distances AB, BC. It can also change the value of the ratio AB/BC. In fact, given any two triples of points A, B, C and A', B', C', each triple on a straight line, it is possible to make two successive projections that send A to A', B to B', and C to C', so you can make the ratio $A'B'/B'C'$ whatever you like.

However, in the case of *four* points on a line, there is a certain quantity, called the cross-ratio of the four points, that always retains its value under projections. Referring to the diagram, the cross-ratio of the four points A, B, C, D is defined by

$$\frac{CA/CB}{DA/DB}.$$

Though length itself is not a notion of projective geometry, the cross-ratio is a projective notion that, at least in the form just described, is based on length. The cross-ratio figures in a number of advanced results of projective geometry.

It is possible to prove Desargues' theorem using techniques of cartesian (Euclidean) geometry, but it is by no means easy. The easiest proof is the one within projective geometry itself.

As is the case with any theorem of projective geometry, if you can prove it for one particular configuration, it will follow for any projection of that configuration. The key step now is to try to prove what, on the face of it, seems to be a harder problem, namely the three-dimensional version of the theorem, where the two given triangles lie in different, nonparallel planes. This more general version is illustrated in the figure on the facing page. The original two-dimensional theorem is, quite clearly, a projection of this three-dimensional version onto a single plane, so proving the more general version will at once yield the original theorem.

Referring to the figure, notice that AB lies in the same plane as $A'B'$. In projective geometry, any two lines in the same plane must meet. Let Q be the point where AB and $A'B'$ meet. Likewise, AC and $A'C'$ intersect at a point R, and BC and $B'C'$ intersect at a point P. Since P, Q, R are on extensions of the sides of ABC and $A'B'C'$, they lie in the same plane as each of these two triangles, and consequently lie on the line of intersection of these two planes. Thus P, Q, R lie on a line, as required.

The artists who developed the principles of perspective realized that in order to create the appro-

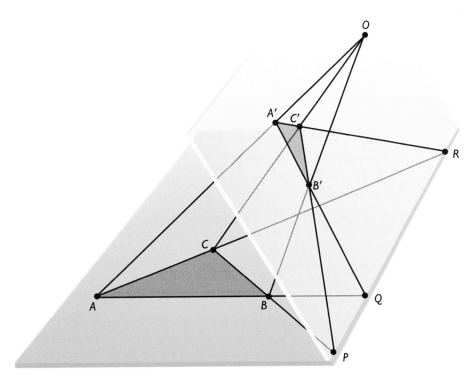

Desargues' theorem in space.

priate impression of depth in a picture, the painter had to establish one or more *points at infinity,* where lines in the picture that correspond to parallels in the scene being painted would meet, if extended. They also realized that these points at infinity must all lie on a single line, called the *line at infinity.* This is illustrated in the figure on the next page.

Similarly, the mathematicians who developed projective geometry realized that it was convenient to introduce points at infinity. Each line in the plane is assumed to have a single 'ideal' point, a 'point at infinity'. Any two parallel lines are assumed to intersect at their common point at infinity. All of the points at infinity are assumed to lie on a single straight line, the 'line at infinity'. This line contains no points other than all the points at infinity.

Notice that only one ideal point is added to each line, not two, as might be suggested by the idea of a line extending to infinity in both directions.

Ideal points and an ideal line were conceived of to avoid issues of parallelism—which is not itself a notion of projective geometry, since projections can destroy parallelism. Because of its Euclidean propensity, the human mind cannot readily visualize parallel lines meeting, so it is not possible to completely visualize the process of adding these ideal points and line. The development has to be axiomatic. With the inclusion of ideal points and an ideal line, projective geometry has the following simple axiomatization:

1. There exist at least one point and one line.

2. If X and Y are distinct points, exactly one line passes through them.

3. If l and m are distinct lines, exactly one point is common to them.

4. There are at least three points on any line.

5. Not all points are on the same line.

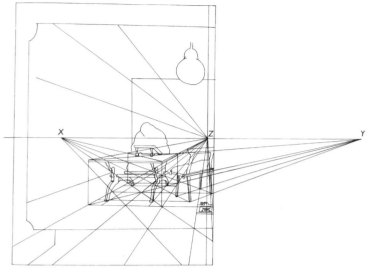

Albrecht Dürer's woodcut *St. Jerome (left)* illustrates multiple-point perspective. An analysis of the perspective *(right)* shows that the three 'points at infinity' (*X, Y,* and *Z*) all lie on a line—the 'line at infinity'.

This axiomatization does not say what a point or a line is, or what it means for a point to lie on a line or for a line to pass through a point. The axioms capture the essential patterns for projective geometry, but do not specify the entities that exhibit those patterns. This is the essence of abstract mathematics.

In this case, one enormous benefit of this level of abstraction is that it practically doubles the number of theorems that can be obtained. Whenever you prove one theorem in projective geometry, a second, so-called dual theorem follows automatically, by what is known as the duality principle. This principle says that if you take any theorem, and everywhere exchange the words 'point' and 'line', and exchange the phrases 'points lying on the same line' with 'lines meeting at a single point', and so on, then the resulting statement will also be a theorem, the *dual* to the first theorem.

The duality principle follows from a symmetry in the axioms. First of all, notice that Axiom 1 is

perfectly symmetric between points and lines, and that Axioms 2 and 3 form a symmetric pair. Axioms 4 and 5 are not in themselves symmetric, but if they are replaced by the statements

4'. There are at least three lines through any point.

5'. Not all lines go through the same point.

then the resulting axiom system is equivalent to the first. Thus, any theorem proved from the axioms will still be true if the notions of point and line are interchanged, together with all associated notions.

For example, in the seventeenth century, Blaise Pascal proved the following theorem:

If the vertices of a hexagon lie alternately on two straight lines, the points where opposite sides meet lie on a straight line.

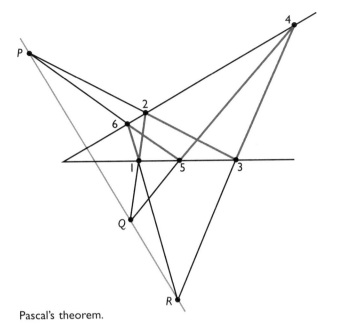

Pascal's theorem.

Brianchon's theorem is illustrated in the second figure on this page. Notice that the hexagon is not regular; it is simply a six-sided figure, with vertices labeled 1, 2, 3, 4, 5, 6. The opposite vertices are 1 and 4, 2 and 5, 3 and 6. Though Brianchon obtained this result by a separate argument, today's geometer would deduce it at once from Pascal's theorem, as the dual.

The dual of Desargues' theorem is its converse:

If, in a plane, two triangles are situated so that the corresponding sides, if extended, intersect in three points that lie on a line, then the straight lines joining corresponding vertices will intersect in a single point.

In this case the duality principle tells us that the converse to Desargues' theorem is true. There is no need for any further proof.

Obviously, the duality principle is only possible if the axioms do not depend on any knowledge of what the words 'point' and 'line' actually mean. When a mathematician is *doing* projective geometry, she may well have a mental picture consisting of regular points and lines, and she may well draw familiar-looking diagrams to help her solve the problem at hand. But those pictures are to help her carry out the reasoning; the mathematics itself requires no such additional information, and, as David

This theorem is illustrated in the first figure on this page. A century later, Charles Julien Brianchon proved this theorem:

If the sides of a hexagon, when extended, pass alternately through two points, the lines joining opposite vertices meet at a single point.

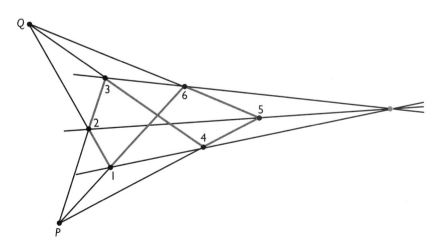

Brianchon's theorem.

Hilbert suggested, would be equally valid if the words 'point' and 'line' were replaced by 'beer mug' and 'table', and the phrases 'lie on a straight line' and 'meet at a single point' were replaced by 'on the same table' and 'supporting the same beer mug', respectively. This particular change results in a system that does not satisfy the axioms for projective geometry *if these words and phrases are taken to have their usual meanings*. For example, a beer mug can rest on only one table. But if the only properties assumed of these words and phrases are those given by the five axioms for projective geometry, then all the theorems of projective geometry will be true of 'beer mugs' and 'tables'.

Dimension

Different geometries represent different ways to capture and study patterns of shape. But there are other aspects of shape, whose study involves tools that, while often closely related to geometry, are generally regarded as not being part of geometry itself. One such is dimension.

The concept of dimension is a fundamental one to human beings. The objects that we encounter each day generally have three dimensions, height, width, and depth. Our eyes work as a coordinated pair in order to achieve a three-dimensional view of our surroundings. The theory of perspective described in the previous section was developed in order to create an illusion of this three-dimensional reality on a two-dimensional surface. Whatever the current theories of physics may tell us about the dimensionality of the universe—and some theories involve a dozen or so dimensions—*our* world, the world of our daily experience, is a three-dimensional one.

But what is dimension? What pattern are we capturing when we speak of dimension? One naive description is in terms of direction, or straight lines, and is illustrated in the figure on this page. A single, straight line represents a single dimension. A second, straight line, perpendicular to the first, in-

If three different colored rods are arranged so that each two meet perpendicular at one end, they will lie in three different dimensions.

dicates a second dimension. A third, straight line, perpendicular to both the first two, indicates a third dimension. At this point the process stops, since we cannot find a fourth straight line perpendicular to each of the three we have so far.

A second explanation of dimension, illustrated on the facing page, is in terms of coordinate systems, as introduced by Descartes. Taken as an axis, a single straight line determines a one-dimensional world. Add a second axis, perpendicular to the first, and the result is a two-dimensional world, or plane. Add a third axis, perpendicular to the first two, and the result is three-dimensional space. Again, the process stops here, since we cannot find a fourth axis perpendicular to the first three.

Both these approaches are unduly restrictive, in that they are too tightly rooted in the Euclidean notion of a straight line. A better way to capture the idea of dimension is in terms of degrees of freedom.

A train traveling on a railway track is moving in one dimension. Though the track itself may curve around, climb, and fall, the train has only one direction of motion. (Reversing is regarded as negative movement forward.) The track is embedded in a three-dimensional world, but the train's world is a one-dimensional one. Relative to an origin, the position of the train at any time may be completely specified by just one parameter, the signed distance from the origin measured along the track. One parameter, one dimension.

A ship on the sea has two degrees of freedom: forward and backward, left and right. The ship therefore moves in two dimensions. Though the surface of the ocean is roughly spherical, curving around the earth, and thus occupies three dimensions in space, the world the ship moves in is two dimensional. Its exact position at any time may be specified using just two parameters, its latitude and its longitude. Two parameters, two dimensions.

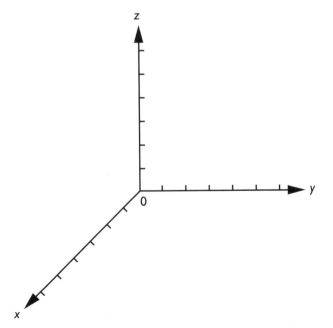

Three perpendicular axes form the basis for Cartesian geometry in three-dimensional space.

An airplane in flight moves in a three-dimensional world. The plane can move forward or (after turning around) backward, left or right, up or down. It has three degrees of freedom, and its position can be uniquely specified using three parameters, latitude, longitude, and altitude. Three parameters, three dimensions.

Degrees of freedom do not have to be spatial, as was the case in the examples just considered. Whenever some system varies according to two or more parameters, if it is possible to alter each parameter without changing the others, then each parameter represents a degree of freedom.

When dimensions are regarded not geometrically but as degrees of freedom, there is no reason to stop at three. For example, the positional status of an airplane in level flight can be specified by five parameters: latitude, longitude, height, speed, and direction of flight. Each of these can be altered independently of all the others. Suppose we wanted to represent the flight in a graphical fashion, as a function of time. The graph would be six dimensional, associating with each point on the time axis a corresponding 'point' in the (five-dimensional) space whose axes represent latitude, longitude, height, speed, and direction. The result would be a time-dependent 'path' or 'curve' of a 'point' moving in a five-dimensional space.

There is no mathematical restriction on the number of possible axes. For any positive integer n, there is an n-dimensional Euclidean space; it is often denoted by E^n. The n coordinate axes for E^n may be labeled $x_1, x_2, \ldots, x_n$. A point in E^n will be of the form $(a_1, a_2, \ldots, a_n)$, where $a_1, \ldots, a_n$ are fixed real numbers. It is then possible to develop geometry algebraically, as a generalization of Descartes' coordinate geometry of the plane.

For example, E^2 is the familiar Euclidean plane. In this geometry, a straight line has an equation of the form

$$x_2 = mx_1 + c,$$

where m and c are constants.

A face-on view of a cube.

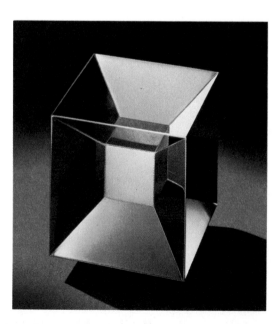

A three-dimensional model representing a face-on view of a hypercube.

In E^3, an equation of the form

$$x_3 = m_1x_1 + m_2x_2 + c$$

represents a plane.

In E^4, an equation of the form

$$x_4 = m_1x_1 + m_2x_2 + m_3x_3 + c$$

represents a "hyperplane," a three-dimensional analogue of a plane.

And so on. The mathematical pattern is quite clear. What changes as you pass from E^3 to E^4 is the ability of human beings to visualize the situation. We can picture E^2 and E^3, but cannot visualize spaces of dimension 4 or higher.

Or can we? Maybe there are ways to obtain some sort of visual impression of objects in four-dimensional space. After all, perspective enables us to create a remarkably good picture of a three-dimensional object from a two-dimensional representation. Perhaps we could construct a three-dimensional 'perspective' model of a four-dimensional object, say a four-dimensional 'hypercube' (the four-dimensional analogue of a cube). Such models have been constructed of wire or metal plates, to better 'see' the entire structure, and show the three-dimensional 'shadow' cast by the four-dimensional object. A photograph of such a model appears at the bottom of this page. This particular model presents a 'cube-on view' of the hypercube, the three-dimensional analogue of the two-dimensional, face-on view of a regular cube, shown in the upper photograph.

In the case of the regular cube, the face-on view is two-dimensional, so the picture you see is the actual view. The nearest face is a square. Because of the effects of perspective, the face farthest away is a smaller square that sits in the middle of the larger, and the remaining faces are all distorted into rhombuses.

Just as the faces of a cube are all squares, so the 'faces' of the four-dimensional hypercube are all cubes. A four-dimensional hypercube is bounded by eight 'cube-faces'. Imagine you have the actual, three-dimensional model shown in the photograph in front of you. The large, outer cube is the cube-face 'nearest' to you. The smaller cube that sits at the center is the cube-face 'farthest away' from you. The remaining cube-faces are all distorted into rhombic pyramids.

Trying to create a mental image of a four-dimensional object from a three-dimensional model (or a two-dimensional picture of such an object) is by no means an easy matter. Even the reconstruction of a mental image of a three-dimensional object from a two-dimensional picture is a highly complex process in which a number of factors come into play, including context, lighting, shading, shadow, texture, and your expectations.

This is the point made by Plato in the allegory of the cave, in the seventh book of *The Republic*. A race of individuals is constrained from birth to live in a cave. All they know of the outside world consists of the colorless shadows cast on the walls of the cave. Though able to gain some sense of the true shape of solid objects, such as urns, by following the changes in the shadows as the objects are rotated, their mental pictures are destined to remain forever impoverished and uncertain.

Plato's cave was brought up to date in 1978, with the production of the movie *The Hypercube: Projections and Slicings,* by Thomas Banchoff and Charles Strauss at Brown University. The movie was created on a computer, and uses color and motion to try to give a sense of the 'true shape' of a four-dimensional hypercube. A frame from this movie is shown on the next page.

With the mental reconstruction of a four-dimensional figure from a three-dimensional model so difficult, the most reliable way to obtain infor-

Many writers and artists have been fascinated by the idea of a fourth dimension. The painting shown is Max Weber's *Interior of the Fourth Dimension* (1913), in the Baltimore Museum of Art.

mation about objects such as the hypercube is to forget trying to picture what is going on and resort to an alternative means of representation: coordinate algebra. Indeed, the *Hypercube* movie was produced in this way. The computer was programmed to perform the necessary algebra, and display the results on the screen in the form of a diagram, using four different colors to represent lines in the four different dimensions.

Using algebra, it is possible to study n-dimensional geometric figures such as n-dimensional hypercubes, n-dimensional hyperspheres, and so forth. It is also possible to use methods of the calculus in order to study motion in n-dimensional space and to compute the hypervolumes of various n-dimensional objects. For example, using integration, the formula for the volume of a four-dimensional hypersphere of radius r can be computed; it is

$$\frac{1}{2}\pi^2 r^4.$$

The mathematical investigation of figures in four or more dimensions turns out to be more than an intellectual exercise having no applications in the real world. The mathematical computer program most widely used in industry is a direct result of such investigations. Called the simplex method, this program tells a manager how to maximize profits—or anything else of interest—in complex situations. Typical industrial processes involve hundreds of parameters: raw materials and components, pricing structures, different markets, a range of personnel, and so forth. Trying to control these parameters so as to maximize profit is a formidable task. The simplex method, invented by the American mathematician George Danzig in 1947, provides a solution based on higher-dimensional geometry. The industrial process is represented as a geometric figure in n-dimensional space, where n is the number of independent parameters involved. In a typical case, this figure will look like an n-dimensional version of a polyhedron; such figures are called *poly-*

topes. The simplex method uses geometric techniques to investigate the polytope, so as to find values of the parameters that result in the maximum profit.

Another use of methods that depend on the geometry of higher-dimensional spaces is in routing telephone calls. In this case, the many different ways to route a call from one part of the county to another can be represented in a geometric fashion as a polytope in an n-dimensional space.

Of course, the computers that carry out these calculations cannot 'see' the geometric figures involved. The computer simply performs the algebraic steps it has been programmed to follow. The n-di-

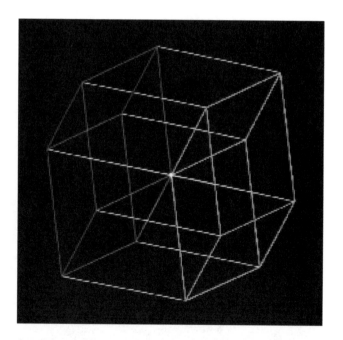

A still frame from the 1978 movie *The Hypercube: Projections and Slicings*, produced on a computer by Thomas Banchoff and Charles Strauss at Brown University. One of the earliest attempts to use computer graphics to visualize complex mathematical objects, this movie won an international award.

mensional geometry is used in devising the program in the first place. Mathematicians may be able to express their thoughts using the language of algebra, but generally they do not think that way. Even a highly trained mathematician may find it hard to follow a long, algebraic argument. But every single one of us is able to manipulate mental pictures and shapes with ease. By translating a complicated problem into geometry, the mathematician is able to take advantage of this fundamental human capability.

As I observed at the start of this chapter, there is a sense in which every one of us is a geometer. By combining geometric ideas with the rigorous methods of algebra in the manner just described, the mathematician has been able to make practical use of that observation.

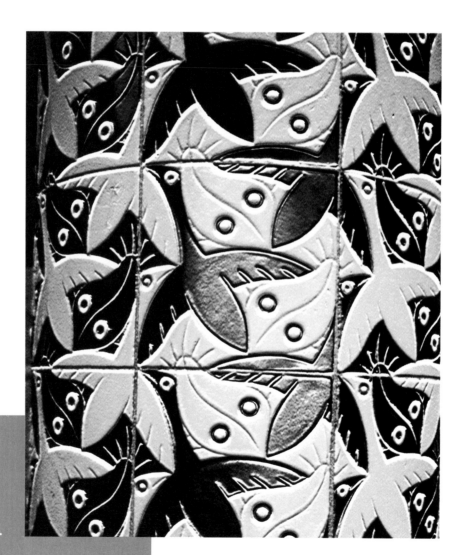

Detail of one
of the columns
designed by
M. C. Escher
in 1968 for a
secondary
school in
Baarn, Holland.

5

Symmetry and Regularity

Geometry sets out to describe some of the visual patterns that we see in the world around us, patterns of shape. But our eyes perceive other patterns, visual patterns not so much of shape per se but of *form*. Patterns of symmetry are an obvious example. The symmetry of a snowflake or a flower is clearly related to the obvious geometric regularity of those objects. The study of symmetry captures one of the deeper, more abstract aspects of shape.

Symmetry Groups

The mathematical study of symmetry is carried out by looking at *transformations* of objects. To the mathematician, a transformation is a special kind of function. Examples of transformations are rotations, translations, reflections, stretchings, or shrinkings of an object. A *symmetry* of some figure is a transformation that leaves the figure invariant, in the sense that, taken as a whole, it looks the same after the transformation as it did before, although individual points of the figure may be moved by the transformation.

An obvious example of a symmetrical figure is the circle. The transformations that leave the circle invariant are rotations about the center (through any angle, in either direction), reflections in any diameter, or any finite combination of rotations and reflections. Of course, a point marked on the circumference may well end up at a different location: a *marked* circle may possess symmetry neither for rotation nor for reflection. But the circle itself, ignoring any marks, does have such symmetry.

The sixfold symmetry of the snowflake. If you rotate a snowflake by any multiple of 60° (one sixth of a complete rotation), it always will look the same.

Given any figure, the *symmetry group* of that figure is the collection of all transformations that leave that figure invariant. A transformation in the symmetry group leaves the figure looking exactly the same, in shape, position, and orientation, as it did before.

The symmetry group of the circle consists of all possible combinations of rotations about the center (through any angle, in either direction) and reflections in any diameter. Invariance of the circle under rotations about the center is referred to as rotational symmetry; invariance with respect to reflection in a diameter is called reflectional symmetry. Both kinds of symmetry are recognizable by sight.

If S and T are any two transformations in the circle's symmetry group, then the result of applying first S and then T is also a member of the symmetry group—since both S and T leave the circle invariant, so does the combined application of both transformations. It is common to denote this double transformation by $T \circ S$. (There is a good reason for the rather perverse looking order here, having to do with an abstract pattern that connects groups and functions, but I shall not go into that connection here.)

This method of combining two transformations to give a third is reminiscent of addition and multiplication, which combine any pair of integers to give a third. To the mathematician, ever on the lookout for patterns and structure, it is natural to see what kind of properties are exhibited by this operation of combining two transformations in the circle's symmetry group to give a third.

First, the operation is associative: if S, T, W are transformations in the symmetry group, then

$$(S \circ T) \circ W = S \circ (T \circ W).$$

In this respect, this new operation is very much like addition and multiplication of integers.

Second, the combination operation has an identity element that leaves unchanged any transformation it is combined with: the 'null rotation', the rotation through angle 0. The null rotation, call it I, can be applied along with any other transformation T, to yield

$$T \circ I = I \circ T = T.$$

The rotation I obviously plays the same role here as the integer 0 does in addition and the integer 1 in multiplication.

Third, every transformation has an inverse: if T is any transformation, there is another transformation S such that

$$T \circ S = S \circ T = I.$$

The inverse of a rotation is a rotation through the same angle in the opposite direction. The inverse of any reflection is that very same reflection. To obtain the inverse for any finite combination of rotations and reflections, you take the combination of backward rotations and re-reflections that exactly undoes its effect: start with the last one, undo it, then undo the previous one, then its predecessor, and so on.

The existence of inverses is a property shared with addition for integers: for every integer m there is an integer n such that

$$m + n = n + m = 0 \text{ (the identity for addition),}$$

namely $n = -m$. The same is not true for multiplication of integers, of course: it is not the case that for every integer m there is an integer n such that

$$m \times n = n \times m = 1 \text{ (the identity for multiplication).}$$

In fact, only for the integers $m = 1$ and $m = -1$ is there another integer n that satisfies the above equation.

To summarize, any two symmetry transformations of a circle can be combined by the combination operation to give a third symmetry transformation, and this operation has the three 'arithmetic' properties associativity, identity, and inverses.

A similar analysis can be carried out for other symmetrical figures. In fact, the properties of sym-metry transformations we have just observed in the case of the circle turn out to be sufficiently common in mathematics to be given a name—indeed, I have already used that name in referring to the 'symmetry *group*'. In general, whenever mathematicians have some set, G, of entities and an operation $*$ that combines any two elements x and y in G to give a further element $x * y$ in G, they call this collection a group if the following three conditions are met:

G1. for all x, y, z in G,
$(x * y) * z = x * (y * z)$;

G2. there is an element e in G such that
$x * e = e * x = x$, for all x in G;

G3. for each element x in G there is an element y in G such that $x * y = y * x = e$, where e is as in condition G2.

Thus, the collection of all symmetry transformations of a circle is a group. In fact, you should have no difficulty in convincing yourself that if G is the collection of all symmetry transformations of *any* figure, and $*$ is the operation of combining two symmetry transformations, then the result is a group.

From the remarks made earlier, it should also be clear that if G is the set of integers and the operation $*$ is addition, then the resulting structure is a group. The same is not true for the integers and multiplication, however. But if G is the set of all rational numbers *apart from zero,* and $*$ is multiplication, then the result is a group.

A different example of a group is provided by the finite arithmetics discussed in Chapter 1. The integers $0, 1, \ldots, n - 1$ with the operation of addition modulo n is a group for any integer n. And if n is a prime number, then the integers $1, 2, \ldots,$ $n - 1$ constitute a group under the operation of multiplication modulo n.

In fact, the three kinds of examples just described barely scratch the surface. The group concept turns out to be ubiquitous in modern mathematics, both pure and applied. Indeed, the notion of a group was first formulated, in the early nineteenth century, not in connection with arithmetic or with symmetry transformations, but as part of an investigation of polynomial equations in algebra. The key ideas may be found in the work of Evariste Galois, described later in this chapter.

The symmetry group of a figure is a mathematical structure that in some sense captures the degree of visual symmetry of that figure. In the case of a circle, the symmetry group is infinite, since there are infinitely many possible angles through which a circle may be rotated and infinitely many possible diameters in which it may be reflected. It is the richness of the circle's group of symmetry transformations that corresponds to the high degree of visual symmetry—the 'perfect symmetry'—that we observe when we look at a circle.

At the other end of the spectrum, a figure that is completely unsymmetric will have a symmetry group that consists only of a single transformation, the identity (or 'do nothing') transformation. It is easy to check that this special case does satisfy the requirements of a group, as does the single integer 0 with the operation of addition.

Before looking at a further example of a group, it is worth spending a few moments reflecting on the three conditions G1, G2, and G3 that determine whether a given collection of entities and an operation constitute a group or not.

The first condition, G1, the associativity condition, is already very familiar to us in the case of the arithmetic operations of addition and multiplication (though not subtraction or division).

Condition G2 asserts the existence of an identity element. Such an element has to be unique. For if e and i both have the property expressed by G2, then, applying this property twice in succession, you would have

$$e = e * i = i,$$

so e and i are in fact one and the same.

This last observation means that there is only one element e that can figure in condition G3. Moreover, for any given element x in G, there is only one element y in G that satisfies the requirement imposed by G3. This is also quite easy to demonstrate. Suppose y and z are both related to x as in G3. That is, suppose that:

(1) $x * y = y * x = e,$

(2) $x * z = z * x = e.$

Then:

$$
\begin{aligned}
y &= y * e   &&\text{(by the property of } e) \\
  &= y * (x * z) &&\text{(by equation (2))} \\
  &= (y * x) * z &&\text{(by G1)} \\
  &= e * z   &&\text{(by equation (1))} \\
  &= z   &&\text{(by the property of } e),
\end{aligned}
$$

so in fact y and z are one and the same. Since there is precisely one y in G related to a given x as in G3, that y may be given a name: it is called the (group) inverse of x, and is often denoted by x^{-1}. And with that, I have just proved a theorem in the mathematical subject known as group theory: the theorem that says that, in any group, every element has a unique inverse. I proved that uniqueness by deducing it logically from the group axioms, the three initial conditions G1, G2, G3.

Though this particular theorem is an extremely simple one, both to state and to prove, it does illustrate the enormous power of abstraction in mathematics. There are many, many examples of groups in mathematics; in writing down the group axioms, mathematicians are capturing a highly abstract pattern that arises in many instances. Having proved, *using only the group axioms,* that group inverses are unique, this fact will apply to every single example of a group. No further work is required. If tomorrow you come across a quite new kind of mathematical structure, and you determine that what you have is a group, you will know at once that every element of your group has a single inverse. In fact, you will know that your newly discovered structure possesses every property that can be established—in abstract form—on the basis of the group axioms alone.

The more examples there are of a given abstract structure, such as a group, the more widespread the applications of any theorems proved about that abstract structure. The cost of this greatly increased efficiency is that one has to learn to work with highly abstract structures, with abstract patterns of abstract entities. In group theory, it does not matter, for the most part, *what* the elements of a group are, or *what* the group operation is. Their nature plays no role. The elements could be numbers, transformations, or other kinds of entities, and the operation could be addition, multiplication, composition of transformations, or whatever. All that matters is that the objects together with the operation satisfy the group axioms G1, G2, and G3.

One final remark concerning the group axioms is in order. In both G2 and G3, the combinations were written two ways. Anyone familiar with the commutative laws of arithmetic might well ask why the axioms were written this way. Why don't mathematician simply write them one way, say

$$x * e = x$$

in G2 and

$$x * y = e$$

in G3, and add one further axiom, the commutative law:

G4. for all x, y in G, $x * y = y * x$.

The answer is that this additional requirement would exclude many of the examples of groups that mathematicians wish to consider.

Though many other symmetry groups do not satisfy the commutativity condition G4, a great many other kinds of groups do. Consequently,

groups that satisfy the additional condition G4 are given a special name: they are called abelian groups, after the Norwegian mathematician Niels Henrik Abel. The study of abelian groups constitutes an important subfield of group theory.

For a further example of a symmetry group, consider the equilateral triangle shown on this page. This figure has precisely six symmetries. There is the identity transformation, I, counterclockwise rotations v and w through 120° and 240°, and reflections x, y, z in the lines X, Y, Z, respectively. (The lines X, Y, Z stay fixed as the triangle moves.) There is no need to list any clockwise rotations, since a clockwise rotation of 120° is equivalent to a counterclockwise rotation of 240° and a clockwise rotation of 240° has the same effect as a counterclockwise rotation of 120°.

There is also no need to include any combinations of these six transformations, since the result of any such combination is equivalent to one of the six given. The table on the next page gives the basic

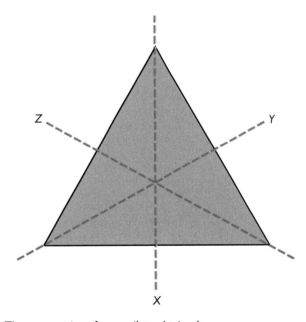

The symmetries of an equilateral triangle.

The Triangle Symmetry Group

○	I	v	w	x	y	z
I	I	v	w	x	y	z
v	v	w	I	z	x	y
w	w	I	v	y	z	x
x	x	y	z	I	v	w
y	y	z	x	w	I	v
z	z	x	y	v	w	I

ushered in by his work. In fact, an entire decade was to go by before the true magnitude of his accomplishment was recognized.

Galois was led to formulate the notion of a group by his attempt to solve a specific problem: that of finding simple, algebraic formulas for the solution of polynomial equations. Every high-school student is familiar with the formula for the solution of a quadratic equation. The roots of the quadratic equation

$$ax^2 + bx + c = 0$$

are given by the formula

$$x = \frac{-b \pm \sqrt{b^2 - 4ac}}{2a}.$$

transformation that results from applying any two basic transformations. To read off the value of the combination $x \circ v$ from the table, look along the row labeled x and locate the entry in the column labeled v, namely y. Thus,

$$x \circ v = y$$

in this group. Again, the result of applying first w and then x, namely the group element $x \circ w$, is z, and the result of applying v twice in succession, namely $v \circ v$, is w. The group table also shows that v and w are mutual inverses and x, y, z are each self-inverse.

Since the combination of any two of the given six transformations is another such transformation, it follows that the same is true for any finite combination. You simply apply the pairing rule successively. For example, the combination $(w \circ x) \circ y$ is equivalent to $y \circ y$, which in turn is equivalent to I.

Evariste Galois

It is to a brilliant young Frenchman by the name of Evariste Galois that the world owes its gratitude for the introduction of the group concept. Killed in a duel on 30 May, 1832, at the age of 21, Galois himself never lived to see the mathematical revolution

Evariste Galois (1811–1832).

Analogous, though more complicated, formulas exist for solving cubic and quartic equations. A cubic equation is one of the form

$$ax^3 + bx^2 + cx + d = 0$$

and a quartic equation is one that has an additional term involving x^4. The formulas for solving these equations are 'analogous' in that they involve nothing more complicated than the evaluation of nth roots, or 'radicals'. Solutions by such formulas are referred to as 'solutions by radicals'.

In 1824, Abel proved that there can be no such formula for a polynomial of the fifth power, known as a quintic polynomial. More precisely, Abel showed that there is no formula that will work for *all* quintic equations. Among polynomials of degree 5 or more, some can be solved by radicals, but others cannot. Galois sought a method to determine, for a given polynomial equation, whether or not that equation was solvable by radicals. The task was as ambitious as his solution was original.

It turned out that whether an equation can be solved by radicals depends on the symmetries of the equation, and, in particular, on the group of those symmetries. Unless you are a budding Galois, or have seen this before, it probably has not occurred to you that equations can have symmetries, or even that they have any kind of 'shape'. Yet they do. The symmetries of an equation are, it is true, highly abstract, but they are symmetries for all that—not visual symmetries but algebraic symmetries. Galois took the familiar notion of a symmetry, formulated an abstract means for describing symmetry (namely, symmetry groups), and then applied that abstract notion of symmetry to algebraic equations. It was as brilliant an example of 'technology transfer' as there has ever been.

To get some idea of Galois' reasoning, take the equation

$$x^4 - 5x^2 + 6 = 0.$$

This equation has four roots, $\sqrt{2}$, $-\sqrt{2}$, $\sqrt{3}$, and $-\sqrt{3}$; substituting any one of them for x will pro-

duce the answer 0. In order to forget the numbers themselves, and concentrate on the algebraic patterns, call these roots a, b, c, and d, respectively. Clearly, a and b form a matching pair, as do c and d. In fact, this similarity goes much deeper than b being equal to $-a$ and d being equal to $-c$: there is an 'algebraic symmetry' between a and b and between c and d. Any polynomial equation (with rational coefficients) satisfied by one or more of a, b, c, and d will also be satisfied if we swap a and b or if we swap c and d, or if we make both swaps at once. If the equation has just one unknown x, swapping, say, a and b will simply amount to replacing a by b in the equation, and vice versa. For instance, a satisfies the equation $x^2 - 2 = 0$, and so does b. In the case of the equation $x + y = 0$, a genuine swap is possible: $x = a$, $y = b$ is a solution and so is $x = b$, $y = a$. For an equation with four unknowns, w, x, y, z, that is solved by a, b, c, d, it is possible to make two genuine swaps at once. Thus, a and b are indistinguishable, and so are c and d. On the other hand, it is easy to distinguish, say, a from c. For example, $a^2 - 2 = 0$ is true, but $c^2 - 2 = 0$ is false.

The permutations of the four roots consisting of swapping a and b, swapping c and d, or performing both swaps together constitute a group known as the Galois group of the original equation. It is the group of symmetries with respect to polynomial equations (with rational coefficients and one or more unknowns) satisfied by the four roots. For obvious reasons, groups that consist of permutations, of which the Galois groups are one kind of example, are referred to as permutation groups.

Galois found a structural condition on groups—that is to say, a property that some groups will possess and others will not—such that the original polynomial equation will have a solution by radicals if, and only if, the Galois group satisfies that structural condition. Moreover, Galois' structural condition depends only on the 'arithmetical' properties of the group. Thus, in principle, whether or not a given equation can be solved by radicals can be determined solely by examining the group table of the Galois group.

To put the historical record straight, Galois did not formulate the concept of an abstract group in the clean, crisp way it was presented above, in terms of the three simple axioms G1, G2, G3. That formulation was the result of efforts by Arthur Cayley and Edward Huntington around the turn of the century. But the essential idea was undoubtedly to be found in Galois' work.

Once Galois' ideas became known, a number of mathematicians developed them further; in particular Augustin Louis Cauchy and Joseph-Louis Lagrange carried out investigations of permutation groups. In 1849, Auguste Bravais used groups of symmetries in three-dimensional space in order to classify the structures of crystals, establishing a close interplay between group theory and crystallography that continues to this day. (See page 164.)

Another major stimulus to the development of group theory was provided in 1872, when Felix Klein, lecturing at Erlangen, in Germany, established what became known as the 'Erlangen program', an attempt, almost entirely successful, to unify geometry as a single discipline. The discussions of the previous chapter should indicate why mathematicians felt the need for such a unification. In the nineteenth century, after the two-thousand-year reign of Euclidean geometry, there had suddenly appeared a whole range of different geometries: Euclidean, Bolyai–Lobachevsky, Riemannian, projective, and several others, including the most recent arrival—and the most difficult to swallow as a 'geometry'—topology, discussed in the next chapter.

Klein proposed that a 'geometry' consists of the studies of those properties of figures that remain invariant under a certain group of transformations (of the plane, of space, or whatever). For example, Euclidean geometry of the plane is the study of those properties of figures that remain invariant under rotations, translations, reflections, and similarities. Thus, two triangles are congruent if one may be transformed into the other by means of a 'Euclidean symmetry', a combination of a translation, a rotation, and possibly also a reflection. (Euclid's definition was that two triangles are congruent if they have corresponding sides equal in length.) Similarly, projective geometry of the plane is the study of those properties of figures that remain invariant under members of the group of projective transformations of the plane. And topology is the study of those properties of figures left unchanged by topological transformations.

With the success of the Erlangen program, a further level of abstract pattern was uncovered: the pattern of different geometries. This highly abstract pattern was described by means of the group-theoretic structures of the groups that determine the geometries.

Sphere Packing

Mathematical patterns can be found everywhere. Just as you can see symmetry when you gaze at a snowflake or a flower, so a trip to your local supermarket can provide another kind of pattern. Take a look at the piles of fruit. How are they arranged? How were they arranged in the crates in which they were shipped? In the case of the fruit pile, the aim is to have an arrangement that is stable; for shipping, the desire is to pack the maximum number of oranges or whatever in the given container. Aside from the rather obvious fact that the absence of sidewalls means that the pile of fruit must have some sort of overall pyramid form, are the two arrangements the same? Do stability and efficiency of packing lead to the same arrangement? If they do, can you explain why?

Like symmetry, the patterns involved in packing objects in an efficient manner can be studied mathematically. The mathematician's version of the supermarket manager's problem of stacking oranges is known as sphere packing. What is the most efficient way to pack identical spheres? Despite a history of investigations into the subject that goes back at least as far as work by Kepler in the seventeenth century, some of the most basic questions remain unanswered to this day.

Thus forewarned, it is perhaps wise to pull back momentarily from the thorny question of sphere

The familiar arrangement of fruit at the market presents mathematicians with a tantalizing problem that remains unsolved to this day. Is the arrangement the most efficient? That is, does it maximize the number of oranges or whatever that can be fitted into the available space?

packing to what must surely be an easier problem: the two-dimensional analogue of circle packing. What is the most efficient way to pack identical circles (or disks) in a given area?

The shape and size of the area to be filled can surely make a difference. So, in order to turn this question into one that is precise and mathematical, you should do what mathematicians always do in such circumstances: fix on one particular case, chosen so as to get to what seems to be the heart of the problem. Since the point at issue is the pattern of the packing, not the shape or size of the container, you should concentrate on the problem of filling *the whole of space*—two-dimensional space in the case of disks, three-dimensional space in the case of spheres. Any answers you obtain in this 'mathematician's idealization' will presumably hold approximately for sufficiently large real-life containers. The larger the container, the better the approximation will be.

The figure on the next page indicates the two most obvious ways to pack disks, called the rectangular and the hexagonal arrangements. How efficient are these two arrangements at packing the maximum number of disks? The decision to concentrate on packing the entire plane means that you have to be a bit careful how you formulate this question in precise terms. The quantity that measures the efficiency of any packing is surely the 'density', the total area or volume of the objects being packed divided by the total area or volume of the container. But when the 'container' is all of the plane or space, calculation of the density in this fashion will produce the nonsensical answer ∞/∞.

The way out of this dilemma can be found in the methods of Chapter 3. The density of packing can be calculated using the same kinds of pattern that gave Newton and Leibniz the key to the calculus: the method of limits. You define the density

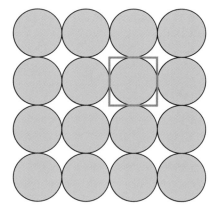

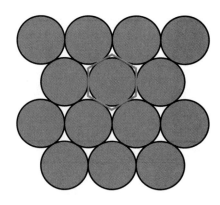

The two regular ways to arrange identical disks to fill the plane. The first is called the 'rectangular' packing, the second is the 'hexagonal' packing. The terminology refers to the figures formed by common tangents to the circles.

Rectangular packing   Hexagonal packing

of a packing arrangement by first calculating the ratio of the total area or volume of the objects packed to the area or volume of the container, for larger and larger *finite* containers. You then compute the limit of those ratios as the containers' boundaries tend to infinity (i.e. get larger without bound). Thus, in the case of disk packing, you can compute the (finite) density ratios for packings that attempt to fill larger and larger square regions whose edges increase without bounds. Of course, as in the case of the calculus, there is no need to compute an endless series of actual, finite ratios; rather you look for the appropriate pattern in the form of a formula and then compute the limit by looking at that pattern.

In the case of disk packing in the plane, this strategy was followed by Kepler, who found that the density of the rectangular packing is $\pi/4$ (approximately 0.785) and that of the hexagonal packing $\pi/2\sqrt{3}$ (approximately 0.907). The hexagonal packing gives the greater density.

Of course, this last conclusion is hardly a surprise: a brief glance at the picture above indicates that the hexagonal packing leaves less space between adjacent disks, and is therefore the more efficient of the two. But it is not quite as obvious that the hexagonal packing is *the* most efficient, that is to say, has a greater density than *any* other packing. The problem with this more general question is that

it asks about *all* possible arrangements of disks, however complex, be they regular or irregular. In fact, so difficult does it seem to resolve the question of what is the most efficient disk packing of all, that it is sensible to look first at a further special case, one that introduces far more pattern than is present in the general case. This is how mathematicians actually did finally arrive at a solution.

In 1831, Gauss showed that the hexagonal packing is the densest among the so-called lattice packings. It was Gauss' concept of a *lattice* that provided the crucial additional structure here, enabling some progress to be made.

In the plane, a lattice is a collection of points arranged at the vertices of a regular, two-dimensional grid. The grid may be square, rectangular, or in the form of identical parallelograms, as illustrated on the facing page. In mathematical terms, the crucial feature of a (planar) lattice is that it has translational invariance, or translational symmetry. This means that certain translations of the plane, without rotation, leave the entire lattice superimposed upon its original position, so that it appears to be unchanged.

A lattice packing of disks is one in which the centers of the disks form a lattice. Such packings are obviously highly ordered. The rectangular and hexagonal disk packings are clearly lattice packings.

Gauss obtained his result that the hexagonal is the most efficient lattice packing by relating lattice packings of disks in the plane to number theory and making use of some number-theoretic work that had been done by Lagrange.

Of course, Gauss' discovery left open the question whether or not the hexagonal arrangement is the most efficient of all disk packings, regular or otherwise. In 1892, Axel Thue announced that he could show that the answer was yes, though it was not until 1910 that he presented a reasonably complete proof.

With the two-dimensional case disposed of, what about the original, three-dimensional sphere-packing problem? Again, it makes sense to do what Gauss did and concentrate first of all on lattice packings, where the centers of the spheres form a three-dimensional lattice, a regular, three-dimensional grid.

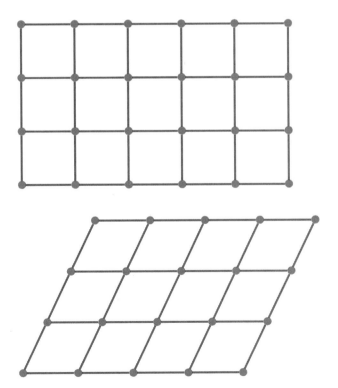

Two planar lattices, one with a square grid (*top*), and one in the form of identical parallelograms (*bottom*).

There are, it turns out, exactly fourteen different kinds of three-dimensional lattice. This result was finally established by the French botanist and physicist Auguste Bravais in 1848, building upon the work of a number of mathematicians. Sometimes referred to as 'Bravais lattices', they are shown on the next page.

One obvious way to arrange spheres in a regular, lattice fashion is to build up the arrangement layer by layer, much as the assistant in the supermarket stacks the oranges. In order to obtain an efficient packing, it seems reasonable to arrange each layer so that the centers of the spheres are one of the two planar lattice formations considered above, the rectangular and the hexagonal. The resulting packings are shown on page 157.

If a rectangular formation is chosen for the layers, there are two ways of stacking the layers one on top of another: so that corresponding spheres are vertically above one another, or staggered so that each sphere in the upper level nestles between four spheres beneath it. (This second alternative is the one used—for stability—in order to stack oranges.) In the former, the centers of the spheres form a cubic lattice; in the latter the centers constitute what is known as a 'face-centered cubic lattice', a cubic lattice in which each cube is 'stood on one corner'.

Alternatively, the layers can consist of hexagonal lattice formations, and again there are two ways in which layers may be stacked one upon another, 'aligned' and 'staggered'. This would appear to give a total of four different three-dimensional lattice packings, but in fact there are only three. Staggered layers of hexagonally packed spheres and staggered layers of squarely packed spheres are equivalent, in that one is just the other viewed from a different angle. You can see this equivalence for yourself quite easily by looking at a pile of oranges arranged in the familiar 'staggered square pyramid' fashion. If you look at one of the slanting faces, you will see one layer of what is a 'staggered hexagonal' packing.

In the third distinct sphere packing, the aligned hexagonal packing, the centers of the spheres constitute what is known as a three-dimensional 'hexagonal lattice'.

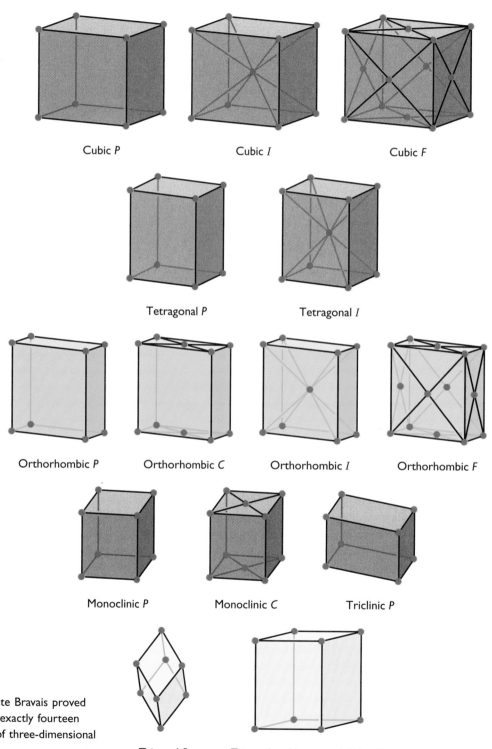

Cubic *P*   Cubic *I*   Cubic *F*

Tetragonal *P*   Tetragonal *I*

Orthorhombic *P*   Orthorhombic *C*   Orthorhombic *I*   Orthorhombic *F*

Monoclinic *P*   Monoclinic *C*   Triclinic *P*

In 1848, Auguste Bravais proved
that there are exactly fourteen
distinct kinds of three-dimensional
lattice.

Trigonal *R*   Trigonal and hexagonal *C* (or *P*)

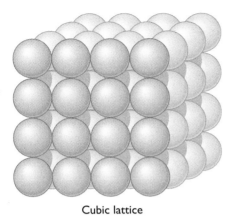

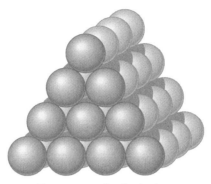

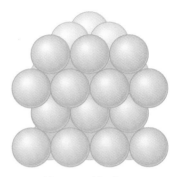

Cubic lattice   Face-centered cubic lattice   Hexagonal lattice

Three different ways to arrange spheres by stacking regular layers. The centers of the spheres constitute a cubic, a face-centered cubic, and a hexagonal lattice, respectively.

Kepler computed the density associated with each of these three lattice packings and obtained the figures $\pi/6$ (approximately 0.5236) for the cubic lattice, $\pi/3\sqrt{3}$ (approximately 0.6046) for the hexagonal lattice, and $\pi/3\sqrt{2}$ (approximately 0.7404) for the face-centered cubic lattice. Thus, the face-centered cubic lattice—the orange-pile arrangement—is easily the most efficient packing of the three. But is it the most efficient of *all* lattice packings? More generally, is it the most efficient of *all* packings, regular or otherwise?

The first of these two questions was answered by Gauss, not long after he solved the analogous problem in two dimensions. Again, he reached his answer by using results from number theory. But the second problem remains unsolved to this day. We simply do not know for sure if the familiar, orange-pile arrangement of spheres is the most efficient of all arrangements.

The orange pile is certainly not the *unique* best packing, since there are nonlattice packings having exactly the same density. It is easy to construct such a nonlattice packing by stacking hexagonal layers as follows. Put down one hexagonal layer. Lay a second on the first in a nestled fashion. Now add a third layer, nestled on the second. There are two distinct ways to do this. In one way, the spheres in the

third layer lie directly above corresponding spheres in the first; in the other way, there is no such alignment. The second of these two alternatives will, if repeated, lead to the face-centered cubic lattice. The first alternative leads to a nonlattice packing that has the same density as the second.

By stacking hexagonal layers and randomly choosing between the two alternatives at each stage, you can obtain a sphere packing that is 'random' in the vertical direction, and yet has exactly the same density as the face-centered cubic arrangement.

Though mathematicians do not know for sure that supermarkets pile oranges in the most efficient fashion, they do know that the arrangement used is very close to the best. It has been proved that no sphere packing can have a density greater than 0.77836.

Kepler's interest in sphere packing, it has to be said, was not driven by an overpowering interest in piles of fruit, but it was, nonetheless, motivated by an equally real phenomenon, that of the shape of a snowflake. Moreover, Kepler's work did involve fruit: not the mundane orange but the—mathematically far more interesting—pomegranate. Honeycombs figured in the investigation as well.

As Kepler observed, although any two snowflakes may differ in many minute ways, they all share the mathematical property of sixfold, or hexagonal, symmetry: if you rotate a snowflake through 60° (one-sixth of a complete rotation), then like a hexagon it will appear unchanged. Why is it, Kepler asked, that all snowflakes have a fundamental hexagonal form?

As usual, Kepler sought the answer in geometry. (Remember, his most famous result was the discovery that the planets follow elliptical orbits. Remember too his fascination with Plato's atomic theory, described on page 113, which was based on the Greeks' five regular solids, and his attempt to describe the solar system in terms of those geometric figures, described on page 114.) His idea was that natural forces impose a regular geometric structure on the growth of such seemingly diverse objects as snowflakes, honeycombs, and pomegranates.

According to Kepler, the key structural notion was that of geometric solids that could be fitted together so as to completely fill space. A natural way to obtain such figures, he suggested, was to start with an arrangement of spheres and imagine that each sphere expands so as to completely fill the intermediate space. Assuming nature always adopts the most efficient means to achieve her ends, the regular patterns of the honeycomb and the pomegranate, and the hexagonal shape of the snowflake, could all be explained by examining efficient packings of spheres and observing the geometric solids they give rise to.

In particular, spheres arranged in a cubic lattice will expand into cubes, spheres arranged in a hexagonal lattice will expand to give hexagonal prisms, and spheres arranged in a face-centered cubic lattice will expand to give Kepler's so-called rhombic dodecahedra, illustrated in the figure to the right. Indeed, in the case of the pomegranate, these theoretical conclusions seem to be substantiated in fact: the seeds in a growing pomegranate are initially spherical, arranged in a face-centered cubic lattice. As the pomegranate grows, the seeds expand until they achieve the form of rhombic dodecahedra that completely fill the internal space.

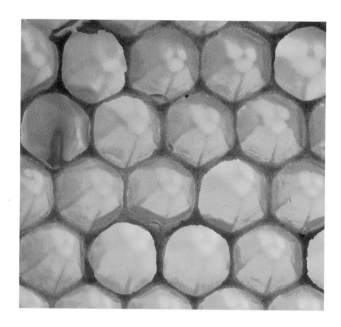

Kepler saw a fundamental, mathematical connection between the formation of a snowflake in the atmosphere, the construction of a honeycomb (*top*) by a family of bees, and the growth of a pomegranate. According to his theory, the regular, symmetric patterns that arise in each case can be described and explained in terms of 'space-filling geometric figures', such as his own discovery, the rhombic dodecahedron (*bottom*), a figure having twelve identical rhombic faces.

In addition to stimulating *mathematical* investigations into sphere packing, Kepler's ideas led to a number of experimental studies of packing. For instance, there is the delightfully titled *Vegetable Staticks,* a 1727 work of an Englishman named Stephen Hales, in which he describes how he filled a pot with peas and compressed them as far as possible, whereupon he observed that the peas had each assumed the shape of a regular dodecahedron. In fact, Hales' claim was presumably exaggerated, since regular dodecahedra do not pack space. However, though a random initial arrangement of spheres will not lead to regular dodecahedra, it will produce various kinds of rhombic shape.

Again, in 1939, two botanists by the name of J. W. Marvin and E. B. Matzke arranged lead shot in a steel cylinder in the familiar orange-stacking manner and then compressed the contents with a piston, obtaining the theoretically predictable rhombic dodecahedra of Kepler. Repeating the experiment with randomly packed shot led to irregular, fourteen-sided figures.

Further experimental work of this nature also demonstrated that random packings are, in general, not as efficient as the face-centered cubic lattice arrangement. The highest density for a random arrangement is around 0.637, as opposed to 0.740 for the familiar orange pile.

Turning to honeycombs, to what do they owe their hexagonal shape? It would be reasonable to suppose that the bees secrete the wax in liquid form, which then forms itself into the observed hexagonal shape under the influence of surface tension. Minimizing surface tension would certainly lead to a hexagonal lattice shape, as proposed by D'Arcy Thompson. Another possibility would be that the bees first hollow out cylinders of wax and then push out the walls of each cylinder until they hit the neighboring cells and fill in the empty space in between. This is what Charles Darwin thought happened.

In fact, neither of these two explanations is correct. The fact is, it is not the laws of inanimate nature that give the honeycomb its elegantly symmetrical shape; rather it is the bees themselves that construct their honeycomb in this fashion. The bees secrete the wax as solid flakes, and construct the honeycomb cell by cell, face by face. The humble bee is in some ways, it seems, a highly skilled geometer, which evolution has equipped for the task of constructing its honeycomb in the form that is mathematically the most efficient.

Finally, what of the snowflake that motivated Kepler to begin his initial study of sphere packing? During the years after Kepler, scientists gradually came to believe that Kepler had been correct, and that the regular, symmetrical shape of crystals reflected a highly ordered internal structure. Finally, in 1915, using the newly developed technique of X-ray diffraction, Lawrence Bragg was able to demonstrate this conclusively. Crystals do indeed consist of identical particles (atoms) arranged in regular lattices.

The snowflake starts out as a tiny, hexagonal crystal 'seed' of ice in the upper atmosphere. As the air currents carry it up and down through altitudes at different temperatures, the crystal grows. The actual pattern that results depends on the particular motions of the growing snowflake through the atmosphere. Since the snowflake is small, the same pattern of growth occurs on all sides, and hence the hexagonal shape of the original seed crystal is preserved. This gives rise to the familiar hexagonal symmetry. (Incidentally, the word 'crystal' comes from the Greek word 'κρισταλλοζ', meaning 'ice'.)

It should be clear by now that a mathematical study of sphere packing can contribute to our understanding of certain phenomena in the world around us. This was, after all, Kepler's reason for commencing such a study. What he most certainly did not anticipate was that the study of sphere packings would have application in the twentieth-century technology of digital communications! Moreover, that application would arise as a result of generalizing the sphere-packing problem to spaces of dimension 4 and more. This surprising, and recent, development is yet another example of a practical application arising from the pure mathematician's search for abstract patterns that exist only in the human mind.

In four and five dimensions, the densest lattice packing of space is the analogue of the face-centered cubic lattice packing, but for dimensions greater than 5, this is no longer the case. The crucial factor is that, as the dimension goes up, there is more and more space between the various hyperspheres. This is illustrated in dramatic fashion in the box on this page. By dimension 8, there is so much free space in the 'face-centered cubic lattice' packing that it is possible to fit a second copy of the same packing into the available gaps without any of the spheres overlapping. The packing that results is the densest lattice packing in eight dimensions. Moreover, certain cross sections of this packing are the densest lattice packings in six and seven dimensions. These results were discovered by H. F. Blichfeldt in 1934.

For the really surprising result, and for the applications to communications technology, you need to go to 24-dimensional space. In 1965, John Leech constructed a remarkable lattice packing in 24-dimensional space based on what is now known as the *Leech lattice*. The Leech lattice, which has deep connections to group theory, gives a sphere packing that is almost certainly the densest lattice packing in 24-dimensional space, with each hypersphere touching 196,560 others. The discovery of the Leech lattice led to a breakthrough in the design of what are known as 'error-detecting codes' and 'error-correcting codes' for data transmission.

Though surprising at first glance, the connection between sphere packing and the design of data codes is a simple one. (In order to illustrate the general idea, I shall, however, simplify matters more than they really are.) Imagine you were faced with designing a means of coding distinct words into digital form in order to transmit messages over some communications network. Suppose you decide to use 8-bit binary strings for your codes, so each word is coded by a string such as (1,1,0,0,0,1,0,1), (0,1,0,1,1,1,0,0), and so on. In transmission of the signal, interference on the channel might cause one or two bits to be miscommunicated, so that an originating string (1,1,1,1,1,1,1,1) arrives as, say, (1,1,1,1,0,1,1,1). In order to be able to detect such

The Sphere-Packing Paradox

The figure below shows four circles, each of radius 1, packed snugly into a 4×4 square 'container'. Adjacent circles just touch each other. Clearly, it is possible to fit a smaller, fifth circle, centered at the origin, so that it just touches the original four circles.

The figure on the facing page illustrates the analogous situation in three dimensions. Eight spheres of radius 1 can be packed tightly into a $4 \times 4 \times 4$ cubic box. A smaller, ninth sphere can obviously be fitted into the center so that it just touches each of the original eight spheres.

Though you cannot visualize it, the same can be done in four, five, or indeed any number of dimensions. For instance, you can pack sixteen four-dimensional hyperspheres of radius 1 into a four-dimensional hypercube that measures $4 \times 4 \times 4 \times 4$, and you can fit an additional hypersphere into the center so that it just touches each of the original hyperspheres. There is an obvious pattern: in dimension n, you can pack 2^n hyperspheres of radius 1 into a hypercube whose edges are all of length 4, and then you can pack an additional hypersphere into the center so that it touches all of them.

Let us see what other patterns can be found. In the original, two-dimensional case, the additional,

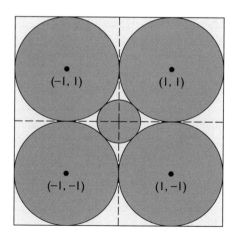

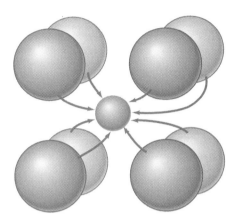

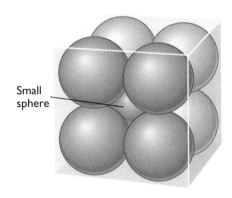

Small sphere

fifth circle is contained inside the bounding outer square, and, in the three-dimensional case, the additional, ninth sphere is contained inside the bounding, cubic container.

Similarly, in dimensions 4, 5, and 6 the additional hypersphere sits inside the hypercube that contains the original hypercubes. The pattern seems so clear that it would be a rare individual who did not assume that the same is true in any number of dimensions. But there is a surprise in store. When you reach dimension 9, something strange happens: the additional nine-dimensional hypersphere actually touches each face of the bounding hypercube; and for dimension 10 and above, the additional hypersphere actually protrudes outside the hypercube. The amount of protrusion increases as the dimension goes up.

This surprising conclusion is easily arrived at in terms of algebraic representations. In the two-dimensional case, what must the radius of the additional circle be in order for it to just touch the original four disks? By the Pythagorean theorem, the distance from the origin to the center of each of the four given circles is $\sqrt{1^2 + 1^2} = \sqrt{2}$. Since each original circle has radius 1, the additional circle must have radius $\sqrt{2} - 1$, which works out to about 0.41. Obviously, an additional circle of radius 0.41 whose center is at the origin will fit easily into the original 4×4 square container.

In the n-dimensional case, by the n-dimensional version of the Pythagorean theorem, the distance from the origin to the center of each of the given hyperspheres is,

$$\sqrt{1^2 + 1^2 + \cdots + 1^2} = \sqrt{n}.$$

So the additional n-dimensional hypersphere must have radius $\sqrt{n} - 1$. In three dimensions, for instance, the additional sphere has radius about 0.73, which again fits easily inside the bounding $4 \times 4 \times 4$ cube. But for $n = 9$, the additional hypersphere will have radius $\sqrt{9} - 1 = 2$, which means it will just touch each face of the bounding hypercube, and for $n > 9$, the radius $\sqrt{n} - 1$ will be greater than 2, and hence the additional hypersphere will protrude outside of the faces of the bounding hypercube.

The point is, the greater the dimension, the more room there is between the original hyperspheres. Even going from two to three dimensions, the additional 'hypersphere' becomes closer to the bounding 'hypercube': the radius increases from 0.41 to 0.73. The eventual protrusion of the additional hypersphere is only surprising because it does not happen until nine dimensions, which is outside of our everyday experience.

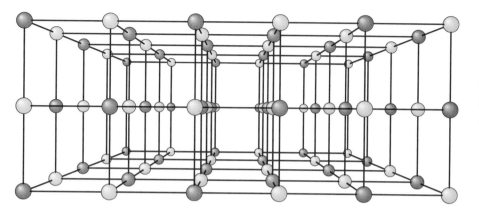

Crystals of common salt (sodium chloride) form perfect cubes. The external form reflects the internal structure of the salt molecule, which is a cubic lattice made up of sodium ions (blue spheres) and chloride ions (orange spheres), alternating in all directions.

The crystal form of the mineral iron pyrite exhibits a distinctive cubic form that reflects its internal molecular lattice structure.

a communication error, it would be sensible to design your coding scheme so that the second of these two strings is not the code of any word, and hence would be recognized as having resulted from a miscommunication. Even better, if you could then recognize what the originating string must most likely have been in order to result in such an arriving string, you could correct the error. In order to achieve these two aims, you need to pick your coding strings so that any two strings that are used as codes differ from each other in at least three binary places. On the other hand, in order to be able to encode all the messages that need to be transmitted, you need to have the largest possible stock of coding strings. If you now view the problem geometrically, what you have is a sphere-packing problem in eight-dimensional space.

To see this, imagine that all possible code words are points in eight-dimensional space. This collection clearly constitutes a 'cubic' lattice, the lattice of all points whose coordinates are 0 or 1. For each string s chosen as a code word, you want to ensure that no other string chosen as a code word differs from s by less than a certain number of bits. In geometric terms, what this amounts to is that no lattice point that lies within a sphere of a particular radius, call it r, centered at s is also a coding string. Maximizing the number of coding strings without having two coding strings that come within a distance r of each other is thus equivalent to finding the densest packing of spheres of radius $r/2$ on that lattice.

And there you have it: from snowflakes and pomegranates to modern telecommunications techniques, by way of the geometry of multidimensional space!

Wallpaper Patterns

In contrast with digital communications, the design of wallpaper patterns might sound frivolous, but if a study of snowflakes and pomegranates can lead to the design of error-correcting codes, there is no telling where an investigation of wallpaper patterns may lead. Certainly, the mathematics of wallpaper patterns turns out to be deep and of considerable intrinsic interest.

The characteristic of a wallpaper pattern of most interest to the mathematician is that it repeats in a regular fashion to completely fill the plane. Thus, the 'wallpaper patterns' the mathematician studies include linoleum floors, patterned cloth, rugs and carpets, and so forth. In these real-life examples, the pattern repeats until the wall, floor, or material ends; the mathematician's pattern stretches out to infinity in every direction.

The mathematical idea behind any wallpaper pattern is that of the *Dirichlet domain* associated with a planar lattice. Given any lattice in the plane and a point of that lattice, the Dirichlet domain of that point consists of the entire region of the plane that is nearer to that point than to any other lattice point. The Dirichlet domains of a lattice provide a 'brick model' of the symmetry of the lattice.

To design a new wallpaper pattern, all you need to do is produce a pattern that will fill one portion of the paper, and then repeat that pattern over all such portions. More precisely, you start off with a lattice grid, fill in one particular Dirichlet domain with your pattern, and then repeat the same pattern in all other Dirichlet domains. Even if the designer does not consciously follow this approach, it is a fact that any wallpaper pattern can be regarded as produced in this manner.

This textile design by William Morris exhibits clear translational symmetry.

There are just five distinct kinds of Dirichlet domain that can arise in the plane, each one either a quadrilateral or a hexagon. The five types are shown on the next page.

There is, of course, no limit to the different kinds of wallpaper pattern that may be designed. But how many of the patterns that you produce in this way are *mathematically* distinct, in the sense of having different symmetry groups? The answer may surprise you: each pattern will be one of only seventeen distinct kinds in terms of symmetry groups. This is because there are exactly seventeen distinct groups that correspond to symmetries of wallpaper patterns. The proof of this fact is quite hard. The

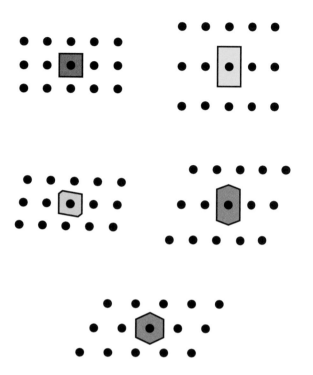

The five distinct kinds of Dirichlet domain for a two-dimensional lattice.

patterns shown on the right illustrate the seventeen different symmetry types. Studies of the various repeating patterns that artists and designers have utilized over the course of history uncover examples of all seventeen possibilities.

The notions of Dirichlet domains and of 'wallpaper patterns' obviously generalize to three dimensions. The fourteen types of three-dimensional lattice illustrated on page 156 give rise to precisely five distinct kinds of Dirichlet domain, shown on the next page. These five figures constitute a collection of solids every bit as fundamental as the five Platonic solids, though they are far less well known.

The five Dirichlet domains give rise to exactly 230 different symmetry groups for three-dimensional 'wallpaper' patterns. Many of these three-dimensional patterns arise in nature, in the structure of crystals. Consequently, the mathematics of symmetry groups plays an important role in crystallography. Indeed, most of the work on classifying the 230 different symmetry groups was carried out by crystallographers, during the late nineteenth century.

Examples of wallpaper patterns of the seventeen distinct possible symmetry groups.

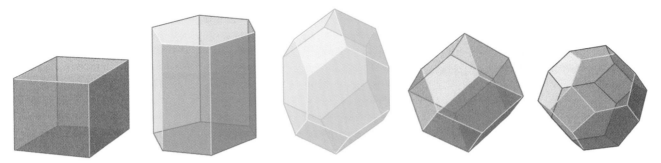

The five distinct kinds of Dirichlet domain for a three-dimensional lattice.

Tiling

While sphere packing investigates the problem of finding the optimal way to arrange a particular shape—a sphere—to achieve the greatest density, the mathematical study of *tilings* looks at a slightly different problem: what *shapes* can be stacked together to fill space completely? This fundamental question is analogous to investigating the manner in which matter splits up into atoms and natural numbers split up into products of primes.

For instance, starting with the two-dimensional case, squares, equilateral triangles, and hexagons can each be arranged to completely fill, or *tile,* all of the plane, as illustrated below. Are these the only regular polygons that can tile the plane? The answer turns out to be yes.

If you allow two or more kinds of tile, but impose the additional requirement that the same array of polygons surrounds each vertex, then there are exactly eight further possibilities, made up of combinations of triangles, squares, hexagons, octagons, and dodecagons, as illustrated on the next page. The regularity of these eleven tilings is sufficiently pleasing to the eye that each will make an attractive pattern for tiling a floor, perhaps to complement one of the seventeen basic wallpaper patterns considered earlier.

If nonregular polygons are allowed, then there is no limit to the number of possible tilings. In particular, any triangle or any quadrilateral can be used to tile the plane. Not every pentagon will work, however; indeed, a tiling consisting of *regular* pentagons will not completely fill the plane, but will leave gaps. On the other hand, any pentagon having a pair of parallel sides will tile the plane. To date, mathematicians have identified fourteen distinct kinds of pentagon that will tile the plane, the

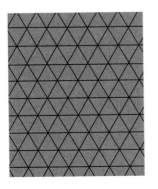
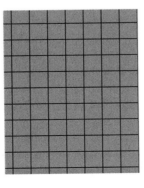
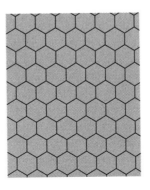

There are three regular polygons that can be arranged to completely fill the plane: the equilateral triangle, the square, and the regular hexagon.

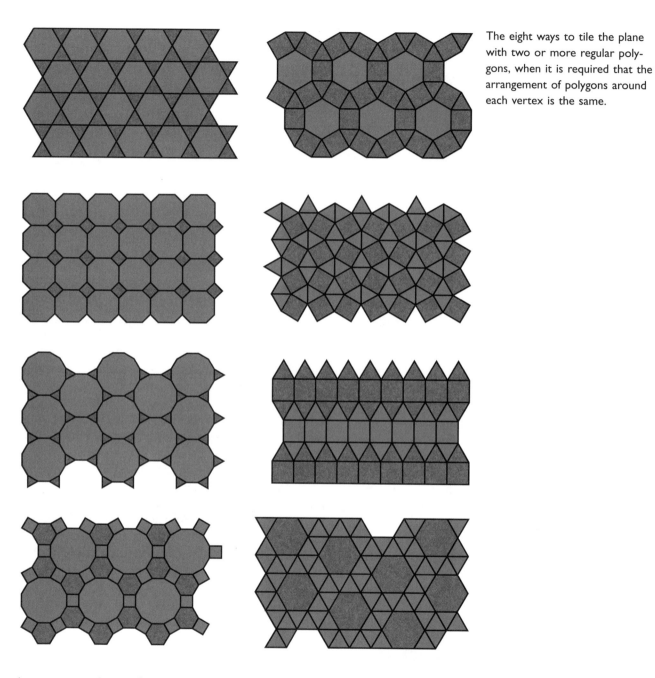

The eight ways to tile the plane with two or more regular polygons, when it is required that the arrangement of polygons around each vertex is the same.

last category being found as recently as 1985, but it is not known if this list is complete. (This finding needs some qualification. It applies only to *convex* pentagons, those in which all corners point outward.)

In the case of hexagons, it was proved back in 1918 that there are precisely three kinds of convex hexagon that will tile the plane; these are shown on the facing page. With hexagons, the possibility of tiling the plane using a single convex polygon comes to an end. No convex polygon having seven or more sides can be arranged to completely fill the plane.

As in the case of sphere packing, where initial studies separated the lattice packings from the irregular ones, it is possible to split the study of tilings into two cases. One case comprises tilings that cover the plane in a repeating (or *periodic*) pattern, having translational symmetry, like those illustrated on pages 165–167. The other case comprises those tilings having no translational symmetry, the *aperiodic* tilings. The distinction between periodic and aperiodic tilings is somewhat analogous to the distinction between rational and irrational real numbers, where the latter have decimal expansions that continue forever, without settling into repeating blocks.

Quite a lot is known about the periodic patterns. In particular, any such tiling of the plane must have as its symmetry group one of the seventeen groups of wallpaper patterns discussed in the previous section. But what of the aperiodic patterns? Indeed, are there any such tilings? Does the plane split up into pieces that have the same shape other than in a periodic fashion? In 1974, British mathematician Roger Penrose gave the—perhaps somewhat surprising—answer yes. Penrose discovered a pair of polygons that could be fitted together to completely fill the plane, but only *aperiodically*, that is to say, without translational symmetry. Penrose's original pair of polygons were not both convex, but two convex polygons that will also tile the plane in an aperiodic fashion were found subsequently, and are shown on the next page. (The new pair is closely related to the original pair found by Penrose.) As was the case with Penrose's own tiles, in order to force the aperiodicity of any tiling using such figures, the edges of the polygons—both rhombuses—have to be assigned a specific direction, and the tiling has to be constructed so that the directions match up along any join. Another way to achieve the same result, but without placing this additional restriction on the tiling procedure itself, is to add wedges and notches to the rhombuses, also illustrated on the next page.

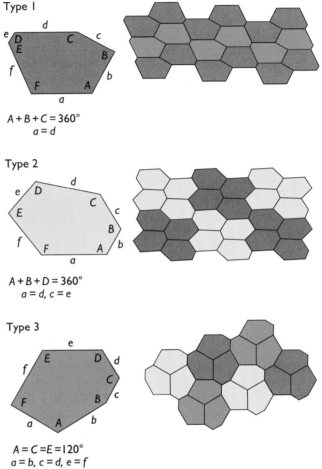

Type 1

$A + B + C = 360°$
$a = d$

Type 2

$A + B + D = 360°$
$a = d, c = e$

Type 3

$A = C = E = 120°$
$a = b, c = d, e = f$

The three ways to tile a plane using a convex hexagon. In order for a hexagon to tile the plane, it must satisfy certain conditions. These are stipulated by the equations given with each figure. For example, in the first figure, the angles A, B, and C must sum 360° and sides a and d must be equal in length. The coloring indicates the basic tiling pattern, which is repeated indefinitely by translation to tile the entire plane.

Readers who have by now grown accustomed to the repeated appearance of various numerical patterns in seemingly very different circumstances may not be too surprised to discover that the golden ratio, ϕ (approximately 1.618), is lurking just beneath the surface here. If the sides of the two rhombuses in the figure above all have length 1, then the long diagonal in the left-hand rhombus is ϕ and the short

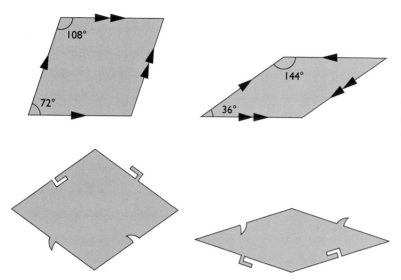

Penrose tilings: by fitting together the two tiles on the top in such a way that the directions of the arrows match at adjoining edges, it is possible to tile the plane, but only in an aperiodic fashion. Adding wedges and notches to the tiles as in the two tiles on the bottom gives polygons that will tile the plane, but only aperiodically, without the need for any restriction on the alignment of tiles.

Part of an (aperiodic) Penrose tiling of the plane, exhibiting the local fivefold symmetry of the tiling.

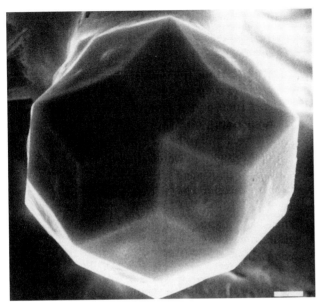 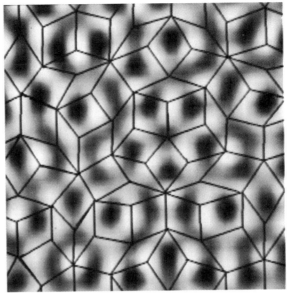

Left: A scanning electron microscope image of the quasicrystal alloy $Al_{5.1}Li_3Cu$. The five-fold symmetry can be seen in the five rhombic faces that meet at a single point in the center of the photograph, forming a starlike shape. *Right*: This image of the quasicrystal material $Al_{65}Co_{20}Cu_{15}$ was obtained with a scanning tunneling microscope and enhanced by a mathematically based process known as Fourier enhancement. The resulting image has been overlaid with a Penrose tiling to display the local fivefold symmetry.

diagonal in the right-hand rhombus is $1/\phi$. When the entire plane is tiled using these two figures, the ratio of the number of fat tiles to thin tiles, computed as a limit, is ϕ.

The figure at the bottom of the facing page illustrates part of a Penrose tiling of the plane. If you look at it for a while, you will notice that small regions of the tiling pattern have a *fivefold symmetry*—if you rotate the region as you would a regular pentagon, the region appears to be unchanged. This fivefold symmetry is, however, strictly *local*. Though various finite regions of the tiling have fivefold symmetry, the entire, infinite tiling does not. Thus, whereas the regular pentagon does not tile the plane, it is nevertheless possible to tile the plane with figures that exhibit local fivefold symmetry. This mathematical discovery, which Penrose made as a

result of a purely recreational investigation, took on added significance in 1984, when crystallographers discovered what are now known as quasicrystals.

The crystallographers observed that certain alloys of aluminum and other elements have molecular structures that exhibit local fivefold symmetry. Since a crystal lattice can have only two-, three-, four-, and six-fold symmetry, these alloys cannot be crystals in the usual sense, hence the new term 'quasicrystal'. In general, a quasicrystal is a material that, while not having the regular lattice structure of ordinary crystals, nevertheless does have its atoms arranged in a highly ordered fashion that exhibits local symmetry.

Whether the structure of any known quasicrystal is that of the Penrose tilings is not clear, and indeed, the study of quasicrystals is still in its infancy, and not without controversy. Nevertheless, the fact that the plane can be tiled in a highly regular, though nonlattice, fashion that exhibits local five-fold symmetry does demonstrate the *possibility* of a mathematical framework that can serve as a basis for understanding these newly discovered materials. Once again we have an example of a development in pure mathematics, discovered for its own sake as a mathematician searches for new patterns, preceding a practical application.

Turning to the question of 'tiling' three-dimensional space, it is clear that the cube will completely fill space. In fact, it is the only regular polyhedron to do so. We have already seen the five *nonregular* polyhedra that fill space in a lattice fashion, on page 165.

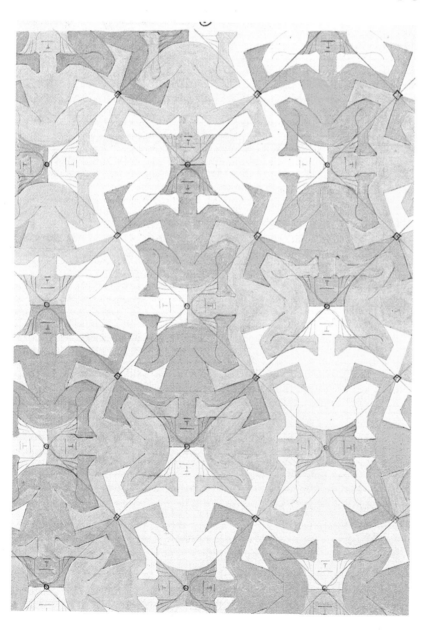

This Escher sketch illustrates how he built up his drawings on a mathematical grid, in this case a square lattice that gives rise to translational symmetry along the lattice lines.

This photograph shows part of an aperiodic tiling of space using cardboard models of the biprism discovered by John Conway in 1993.

Three-dimensional analogues of the aperiodic, Penrose tilings have been known for several years. Penrose himself found a pair of *rhombohedra* (squashed cubes) which, when stacked according to various alignment restrictions (much as in the case of the two-dimensional analogue), fill all of three-dimensional space in an aperiodic fashion.

In 1993, to the surprise of most mathematicians, a single, convex polyhedron was discovered that could be arranged to fill space completely, but only in an aperiodic fashion. Credit for the discovery of this remarkable solid goes to British mathematician John Horton Conway. Conway's new solid, shown on this page, is what is known as a 'biprism'; that is, it is formed from two slant triangular prisms fused together. Its faces consist of four congruent triangles and four congruent parallelograms. In order to fill space using this polyhedron, you proceed layer by layer. Each layer is periodic, but in order to lay a second layer on top of a first, the second must be rotated by a fixed, irrational angle, and this 'twist' ensures that the tiling is aperiodic in the 'ver-

tical' direction. (An aperiodic tiling of space using *non*convex polyhedra had previously been achieved in this fashion by Austrian mathematician Peter Schmitt.)

It is not known if there is a single, two-dimensional figure that will tile the plane (only) aperiodically. So, in the case of aperiodic tilings, our knowledge of the three-dimensional case exceeds that for two dimensions.

Despite its relevance to designers and its interest for the occasional recreational mathematician, the study of tilings remained a relatively obscure branch of mathematics until the last twenty years or so. It is now a thriving area of research that has found a number of surprising applications in other parts of mathematics, as well as to such tasks as the distribution of supplies and the design of circuits. With that increased interest, mathematicians are finding that there is much about tiling that they do not understand, providing still further testament to the fact that deep and challenging mathematical problems arise in all aspects of our lives.

Mauritz Escher,
Three Spheres,
1945 (woodcut).

6

Position

T he London Underground map, shown on the next page, was first drawn in 1931. Apart from a number of subsequent additions as the underground system grew, the map remains largely in its original form. Its longevity is a testament to its utility and its aesthetic appeal. And yet, in geometric terms, it is hopelessly inaccurate. It is certainly not drawn to scale, and if you try to superimpose it on a standard map of London, you will find that the positions of the stations on the Undergound map are not at all correct. What *is* correct is the representation of the *network*. The map tells you what subway line to take to get from point A to point B, and where to change lines if necessary. And this, after all, is the only thing that matters to the Underground traveler—one hardly takes the underground for the view en route! The crucial point is that in this one respect the Underground map is accurate, completely so. It therefore succeeds in capturing an important pattern in the geography of the London Underground system. That pattern is what mathematicians call a *topological* pattern.

In two dimensions, the mathematical discipline known as *topology* is sometimes referred to as 'rubber sheet geometry', since it

The London Underground Map

The London Underground map was designed in 1931 by Henry C. Beck, a twenty-nine-year-old, temporary draughtsman working for the London Underground Group; but it took two years of persistent efforts by Beck before his now-familiar map was accepted for publication. Even then, the Underground Publicity Department produced the map only in small numbers. Their fear was that the map's total abandonment of geographical accuracy would render it incomprehensible to the majority of Underground travelers. They were wrong. The public loved it, and by the end of its first year in use, a larger version was posted all over the system. Without the need for any explanation or training, the general public not only coped easily with their first explicit encounter with a genuinely *topological* representation of the Underground network, they recognized at once its advantages over the more familiar geometric depictions.

studies properties of figures that are unchanged by stretching or twisting the surface on which the figures are drawn. The topological nature of the Underground map is illustrated with great frequency in London these days due to the manufacture and sale of souvenir T-shirts bearing the map on the chest. Such maps continue to provide a completely reliable guide to underground travel whatever the shape of the body they adorn, though propriety dictates that the student of topology not take this particular aspect of the study too far.

Topology is one of the most fundamental branches of present-day mathematics, with ramifications in many other areas of mathematics, as well as in physics and other scientific disciplines. The name 'topology' comes from the Greek τοπος λογος [*topos logos*], meaning 'the study of position'.

The Königsberg Bridges

As is so often the case in mathematics, the broad-ranging subject known as topology has its origins in a seemingly simple, recreational problem, the Königsberg bridges problem.

The City of Königsberg (now called Kaliningrad), which was located on the River Pregel in East Prussia, had two islands, joined together by a bridge. As shown below, one island was connected to each bank by a single bridge, the other island had two bridges to each bank. It was the habit of the more energetic citizens of Königsberg to go for a long family walk each Sunday, and, naturally enough, their path would often take them over several of the bridges. An obvious question was whether there was a route that traversed each bridge exactly once.

Euler solved the problem in 1735. He realized that the exact layout of the islands and bridges was irrelevant. What was important was the way the bridges connect, that is to say, the *network* formed by the bridges, illustrated in the figure at the top of this page. The actual layout of the river, islands, and bridges—that is to say, the geometry of the problem—is irrelevant. In Euler's network, the bridges are represented by *edges*, the two banks and the two islands are represented by *vertices*. In terms of the network, the problem asks if there is a path that follows each edge exactly once.

Euler argued as follows. Consider the vertices of the network. Any vertex that is not a starting or a

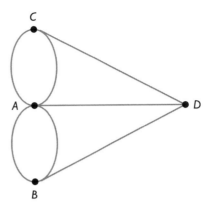

Euler's solution to the problem of the Königsberg bridges depended on his first observing that the crucial feature was the network formed by the bridges.

finishing point of such a path must have an even number of edges meeting there, since those edges can be paired off into 'path-in–path-out' pairs. But, in the bridges network, *all four* vertices have an odd number of vertices that meet there. Hence there can be no such path. In consequence, there can be no tour of the bridges of Königsberg that crosses each bridge exactly once.

Euler was able to solve the Königsberg bridges problem by realizing that it had almost nothing to do with geometry. Each island and each bank could be regarded as a single point, and what counted was the way these points were connected—not the

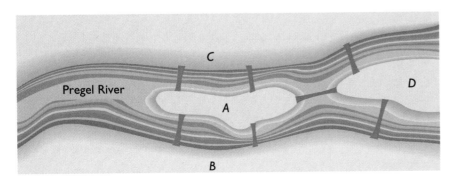

Is it possible to make a tour that traverses each of the seven bridges exactly once? In 1735, Leonhard Euler proved that it is not.

length or shape of the connections, but which points were connected to which others.

This independence from geometry is the essence of topology. Euler's solution to the problem of the Königsberg bridges gave rise to a substantial branch of topology known as network theory. Network theory has many present-day applications; the analysis of communications networks and the design of computer circuits are two very obvious examples.

Networks

The mathematician's definition of a network is very general. Take any collection of points (called vertices of the network) and connect some of them together by lines (called edges of the network). The shape of the edges is unimportant, though no two edges may intersect (except at their ends) and no edge may intersect itself or form a closed loop. (You may, however, form a closed circuit by joining together two or more distinct edges at their ends.) In the case of networks in the plane, or on some other two-dimensional surface such as the surface of a sphere, the nonintersection requirement is fairly restrictive. These are the kinds of networks I shall consider here. For networks in three-dimensional space, edges can pass over each other without problem.

One additional restriction is that any network must be 'connected'; that is to say, it must be pos-

sible to go from any vertex to any other vertex by tracing a path along the edges. A number of networks are illustrated on this page.

The study of networks in the plane leads to some surprising results. One such is *Euler's formula*, discovered by Euler in 1751. For networks in the plane or on any other two-dimensional surface, the edges of a network divide the surface into distinct regions called the *faces* of the network. Take any network and calculate the number of vertices (call that number V), the number of edges (call it E), and the number of faces (call it F). If you now compute the value of the sum

$$V - E + F,$$

you will find that the answer is 1. Always.

This result is clearly quite remarkable: no matter how simple or how complex a network you draw, and no matter how many edges your network has, the above sum always works out to be 1. It is not hard to prove this fact. Given any network, start erasing edges and vertices from the outside, working your way inward, as illustrated in the figure on the facing page. Take care to ensure that the network remains connected at each stage. Removing one outer edge (but not the vertices at either end) reduces the number E by 1, leaves V unchanged, and decreases F by 1. Thus the net effect on the value of $V - E + F$ is nil, since the reductions in E and F cancel each other out. Whenever you have a

$V = 7, E = 12, F = 6$

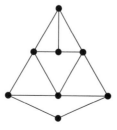

$V = 8, E = 13, F = 6$

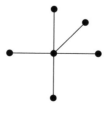

$V = 6, E = 5, F = 0$

Euler proved that, for any network drawn on the plane, the sum $V - E + F$ is always equal to 1, where V is the number of vertices, E is the number of edges, and F is the number of faces enclosed by edges of the network.

Removal of an outer edge

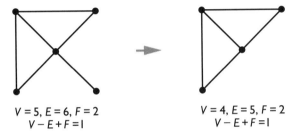

$V = 5, E = 6, F = 2$
$V - E + F = 1$

$V = 5, E = 5, F = 1$
$V - E + F = 1$

Removal of a dangling edge

$V = 5, E = 6, F = 2$
$V - E + F = 1$

$V = 4, E = 5, F = 2$
$V - E + F = 1$

The two key steps in the proof of Euler's formula for networks in the plane. See the text for details.

'dangling edge', that is to say, an edge terminating at a vertex to which no other edge is connected, remove that edge and the free vertex. (If both ends are free, remove one vertex and leave the other as an isolated point. The choice of which to remove and which to leave is up to you.) This reduces both V and E by 1, and leaves F unchanged, so again the value of $V - E + F$ is not altered by this procedure.

If you continue in this fashion, removing outer edges and pairs of dangling edges and vertices one after another, you will eventually be left with a single, isolated vertex. For this, the simplest of all networks, the sum $V - E + F$ is obviously 1. But the value of this sum did not change at any stage of the reduction procedure, so its value at the end will be exactly the same as at the beginning. So the initial value of $V - E + F$ must have been 1. And there you have your proof!

The fact that the sum $V - E + F$ is equal to 1 for all networks is analogous to, say, the fact that

the sum of the angles of any triangle is 180°. Different triangles can have different angles and sides of different lengths, but the sum of the angles is always 180°; analogously, different networks can have different numbers of vertices and edges, but the sum $V - E + F$ is always 1. But, whereas the triangle angle sum depends crucially on shape, and is thus a fact of geometry, Euler's $V - E + F$ result is completely independent of shape. The lines of the network may be straight or crooked, and the surface on which the network is drawn may be flat or undulating, or even folded. And if the network is drawn on a material that can be stretched or shrunk, neither of these manipulations will affect the result either. All of this is surely self-evident. Each of these manipulations will affect geometric facts, so the $V - E + F$ result is not a geometric fact. A fact about figures that does not depend upon bending or twisting or stretching is known as a *topological* fact.

What happens if the network is drawn not on a plane but on the surface of a sphere? You can try this by using a felt-tipped pen to draw a network on an orange. Provided your network covers the entire surface (that is, provided you are not using just part of the surface, as if it were simply a curved sheet), then you will find that the sum $V - E + F$ turns out to be not 1 but 2. So, whereas bending, twisting, stretching, and shrinking a plane sheet cannot effect the value of this sum, replacing the sheet by a sphere does make a difference. On the other hand, it is easy to convince yourself that bending, twisting, stretching, or shrinking the sphere will not change the value of $V - E + F$ in that case either; it will always be 2. (You can check this by drawing a network on a balloon.)

Proving that the value of quantity $V - E + F$ for networks on a sphere is always 2 is no more difficult than proving the analogous result for the plane; the same kind of argument can be used. But there is another way to prove this fact. It is a *consequence* of Euler's formula for networks in the plane, and can be deduced from that previous result by topological reasoning (as opposed to, say, geometric reasoning, which is what Euclid used to prove the theorems in *Elements*).

To begin the proof, imagine your network drawn on a sphere which is perfectly stretchable. (No such material is known, but never mind, the mathematician's patterns are always in the mind.) Remove one complete face. Now stretch the edges surrounding the missing face so that the entire remaining surface is flattened out into a plane, as illustrated on this page. This stretching process clearly will not affect the value of $V - E + F$ for the network, since it will not change any of V, E, or F.

When the stretching is complete, you have a network on the plane. But we already know that, in this case, the value of $V - E + F$ is 1. In going from the original network on the sphere to the network on the plane, all you did was remove one face; the vertices and edges were left intact. So that initial move caused a reduction in the value of $V - E + F$ of 1. Hence, the original value of $V - E + F$ must have been exactly 1 more than the value you finished with for the plane network, which means that the original value for a network drawn on a sphere was 2.

Closely related to this result about networks on spheres is the following fact about polyhedra. If V denotes the number of vertices of a polyhedron, E the number of edges, and F the number of faces, then it is always the case that

$$V - E + F = 2.$$

The 'topological' proof is quite obvious. Imagine that the polyhedron is 'blown-up' into the form of a sphere, with lines on the sphere denoting the position of the former edges of the polyhedron. The result is a network on the sphere, having the same values for V, E, F as the original polyhedron.

In fact, it was in this form, as a result about polyhedra, that the identity $V - E + F = 2$ was first observed, by René Descartes in 1639. However, Descartes did not know how to prove such a result. When applied to polyhedra, the result is known as 'Euler's polyhedra formula'.

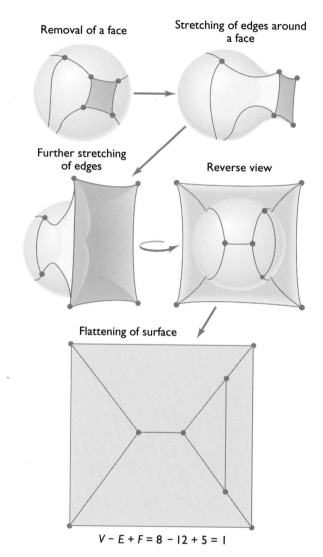

Removal of a face

Stretching of edges around a face

Further stretching of edges

Reverse view

Flattening of surface

$$V - E + F = 8 - 12 + 5 = 1$$

By removing a single face and stretching the remaining surface to be planar, a network on a sphere can be transformed into a network on the plane that has the same number of vertices and edges, but one less face.

Topology

Euler was not the only eighteenth-century mathematician to investigate topological phenomena; both Cauchy and Gauss realized that figures had properties of 'form' more abstract than the patterns of geometry. But it was Gauss' student Augustus

Möbius who really set in motion the mathematical discipline now known as topology, by giving a precise definition of a *topological transformation*. Given Möbius' definition, topology is the study of the properties of figures left invariant by topological transformations.

A topological transformation is a transformation from one figure into another such that any two points close together in the original figure remain close together in the transformed figure. It takes a bit of work to make precise just what is meant by the phrase 'close together' in this definition; in particular, stretching is allowed, though this operation clearly increases the distance between points. But the intuition is clear enough. The most significant manipulation that is prohibited by this definition is cutting or tearing, except in the case where a figure is cut in order to perform some manipulation and then 'glued' back together again so that points on either side of the cut that were originally close together are close together when the transformation is complete.

Most of the early work in topology was directed toward the two-dimensional case, the topological study of *surfaces*. A particularly fascinating discovery was made early on by Möbius together with another of Gauss' students by the name of Johann List-

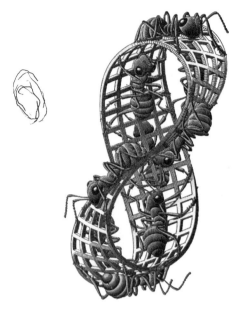

Escher's conceptualization of a Möbius band. The ants are destined to continue walking forever, covering the same ground over and over again.

ing. They found that it was possible to have 'one-sided' surfaces. If you take a long strip of paper, say one inch wide by ten inches long, give it a single half-twist, and then glue the two free ends together, the result is a surface having only one side, and known nowadays as a Möbius band. A surface of this type is illustrated on this page. If you try to color just one side of a Möbius band, you will find that you end up coloring what you might have thought were 'both sides' of the figure.

This, at least, is the way mathematicians often present the Möbius band to children or to beginning students of topology. As is often the case, however, the true story is more subtle.

First of all, *mathematical* surfaces do not have 'sides'. The notion of a 'side' is something that results from observing a surface from the surrounding three-dimensional space. To a two-dimensional creature constrained to live 'in' a surface, the notion of a side makes no sense at all, just as it makes no sense

The Möbius band is constructed by giving a strip of paper a single half-twist and then attaching the two free ends to form a closed loop. It is popularly described as being a surface having just one side and one edge.

to us to talk about our own, three-dimensional world having sides. (Viewed from four-dimensional space-time, our world does have sides, namely past and future. But these notions only make sense when you include the additional dimension of time: past and future refer to the position of the world in time.)

Since mathematical surfaces do not have sides, it follows that you cannot draw a figure 'on one side' of a surface. Mathematical surfaces have no thickness; a network may be *in* a surface, but not on it. (Mind you, mathematicians often do speak of 'drawing' a network 'on' a surface. After all, this terminology is appropriate for the kinds of nonmathematical surfaces we encounter in our daily lives. But when carrying out a mathematical analysis, the mathematician will be careful to treat a network or other figure as being *in* the surface, not 'on' it.)

Thus, it is not mathematically correct to say that the Möbius band has 'one side'. What is it then that makes the Möbius band so different from a regular, cylindrical band, formed by looping a strip of paper without giving it an initial half-twist? The answer is that the ordinary band is *orientable*, whereas the Möbius band is *nonorientable*. The mathematical notion of orientability is a genuine property that mathematical surfaces may or may not have. Intuitively, orientability means that there are distinct notions of clockwise and counterclockwise, or that left- and right-handedness are different.

To obtain an initial grasp of this abstract notion, imagine two bands, a simple, cylindrical band and a Möbius band, constructed from transparent film, such as clear photographic film or the film used for overhead-projector transparencies. (Even better, make two such bands and follow the discussion physically.) Draw a small circle on each band, and add an arrowhead to indicate the 'clockwise' direction. Of course, whether you call this direction 'clockwise' or 'counterclockwise' depends on how you view it from the surrounding three-dimensional space. Using transparent film means you can see your circle from both 'sides'; the use of this material thus provides a better model of a genuine, *mathematical* surface than does a sheet of opaque paper.

Imagine sliding your circle all the way around the band until it returns to its original position. To simulate this action on your model, you could start with your circle and proceed around the band, in one direction, making successive copies of the circle, together with the directional arrowhead, at some fixed, small distance apart, as illustrated in the drawing on this page. When you have gone completely around the band, you will find yourself making your copy on the other physical side of the film from the original circle. Indeed, you can superimpose the new circle over the original one. But the directional arrowhead, copied from the previous circle, will point in the opposite direction to the original one. In the course of moving around the band, the *orientation* of the circle changes. But since the circle remained 'in' the surface the whole time, and was simply moved around, this result means that, *for this particular surface*, there is no such thing as 'clockwise' or 'counterclockwise'; such a notion simply makes no sense.

On the other hand, if you repeat the procedure for the cylindrical band, the result is quite different. When you complete the circumvention of the band, and the circle has returned to its original position, the directional arrowhead points in the same direction as before. For figures drawn 'in' the cylin-

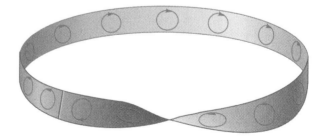

The genuine topological property of the Möbius band that corresponds to its 'having only one side' is that it is nonorientable. It is possible to transform a clockwise direction into a counterclockwise direction by simply moving once round the band.

drical band, you cannot change the orientation by moving the figure about.

The abstract notion of orientability can also be understood in terms of 'handedness'. If you draw the outline of a human hand on a (transparent) Möbius band and transport it once around the band, then you will find that it changes 'handedness': what you (looking at the band from the surrounding three-dimensional space) perhaps thought of as left- and right-handedness have been interchanged. This means that, for the Möbius band, there is no notion of left- or right-handedness.

Orientability is a genuine topological property of surfaces. Since the cylindrical band is orientable and the Möbius band is nonorientable, these two surfaces must be topologically distinct. Hence, it cannot be possible to transform a cylindrical band into a Möbius band by means of a topological transformation. This surely accords with our intuitions. The only obvious way to transform physically a Möbius band into a cylindrical band is by cutting the band, removing the half-twist, and then reattaching the two cut ends. But the act of removing the half-twist means that when the two free ends are reattached, points originally close together that were separated by the cut are no longer close together, and thus the transformation is not a topological one.

The distinction between properties of a surface and properties of the surrounding space is dramatically illustrated by constructing a third band, differing from the Möbius band in that you give the strip a full twist (rather than a half-twist) prior to attaching the free ends. This new strip is topologically equivalent to the cylindrical band, since one can be transformed into the other by cutting the twisted band, undoing the twist, and then reattaching the two cut ends. Parts of the band on either side of the cutting line that are close together initially will be close together after this operation is complete, so this is a genuine topological transformation.

Now perform the following operation on your three bands. Take a pair of scissors and cut along

the center, all the way around the band. The result in each of the three cases turns out to be quite different and, if you have not seen it before, extremely surprising. In the case of the cylindrical band, you end up with two separate cylindrical bands the same length as the original one. Cutting the Möbius band produces a single, full-twist band, twice as long as the original band. Cutting the full-twist band produces two interlocked full-twist bands, each the same length as the original one. Whereas you might assume that the difference in outcome of cutting the cylindrical band and the Möbius band is attributable to topological differences between these surfaces, the difference in outcomes between cutting the cylindrical band and the full-twist band cannot be explained in this way, since the two are topologically equivalent. The different outcomes arise from the way the bands are embedded in the surrounding three-dimensional space.

You might also try a further experiment. Take each of your three bands, and cut along the length of the band much as before, only start your cut not in the middle of the band but one-third of the way from one edge. How does the outcome this time compare with the outcome in the previous case?

Orientability is not the only topological property that can be used to distinguish between surfaces. The number of edges is another topological feature of a surface. The sphere has no edges, the Möbius band has one edge, and the cylindrical band has two edges. (You can check that the Möbius band has only one edge by running a colored crayon along the edge.) Thus, the difference in the number of edges is another topological property that distinguishes the Möbius band from the cylindrical band. On the other hand, in terms of edges, the Möbius band is the same as a plane disk, which also has one edge. In this case, orientability is a topological property that distinguishes the two surfaces: the disk is orientable, the Möbius band is not.

What about a disk with a hole somewhere in the middle? This surface is orientable, and has two edges, just like the cylindrical band. And in fact the two surfaces are topologically equivalent; it is easy to see how to turn a cylindrical band into a disk with a hole in the middle by (mathematical) stretching and flattening out.

So far, all of this is certainly of interest, not to say entertaining. But the topological properties of surfaces are of far more than intrinsic interest. As is generally the case, when mathematicians discover a truly fundamental kind of pattern, it turns out to have widespread application, and this is particularly true of topological patterns. Historically, what really established the new discipline of topology as a central plank of modern mathematics was the development of complex analysis, the extension of the methods of the differential calculus from the real numbers to the complex numbers, which we explored at the end of Chapter 3.

Since the real numbers lie on a one-dimensional line, a function from real numbers to real numbers can be represented by a line in the plane—a *graph* of the function. But complex numbers are two-dimensional, and so a function from complex numbers to complex numbers is represented not by a line but by a surface. The simplest case to imagine is that of a continuous function from complex numbers to real numbers. Here, the 'graph' of the function is a surface in three-dimensional space, where the real value of the function at any complex point is regarded as a 'height' above or below the complex plane. It was Bernhard Riemann's use of surfaces in complex analysis, around the turn of the century, that brought the study of the topological properties of surfaces to the forefront of mathematics, where it has remained ever since.

Classification of Surfaces

With the importance of the study of surfaces established, mathematicians needed a reliable means for the topological *classification* of surfaces: what features of surfaces suffice to classify surfaces topologically, so that any two surfaces that are topologically equivalent share those features, and any two surfaces that are not topologically equivalent may be distinguished by one or more of those features? In Eu-

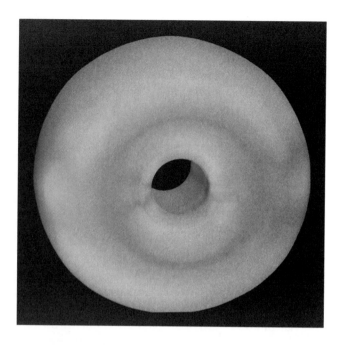

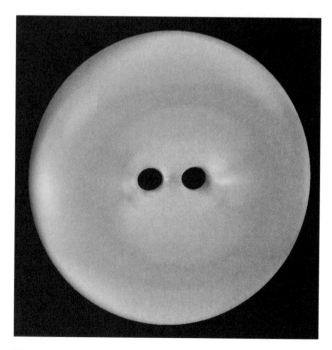

The torus and the double torus. Each is a closed surface. As a surface, neither of these figures has a hole; the holes are in the surrounding three-dimensional space, not in the surface.

clidean geometry, for example, polygons may be classified according to the number of their edges, the lengths of those edges, and the angles between them.

The number of edges is one feature used to classify surfaces topologically. Orientability is another. These suffice to distinguish, say, spheres, cylinders, and Möbius bands. But they do not distinguish between the torus (illustrated on the facing page) and the sphere, both of which have no edges and are orientable. Of course, it is tempting to say that the torus has a hole in the middle, and the sphere does not. The problem is that the hole is not a part of the surface itself, any more than sidedness is. The hole of the torus is a feature of the manner in which the surface sits in three-dimensional space. A small, two-dimensional creature constrained to live *in* the surface of the torus would never encounter the hole. The question for the would-be classifier of surfaces, then, is to find some topological property of the surface that such a creature might be able to recognize and that is different for the torus and the sphere.

What topological properties are there, other than the number of edges and orientability? One possibility is suggested by Euler's result about the value of $V - E + F$ for networks in a surface. The values of V, E, and F for a given network are unchanged if the surface in which the network is drawn is subjected to a topological transformation. In addition, the value of the quantity $V - E + F$ does not depend on the actual network drawn (at least in the case of the plane or the sphere). So perhaps this quantity $V - E + F$ is a *topological invariant* of surfaces in general.

And indeed it is. The kind of reduction argument Euler used to establish the constancy of $V - E + F$ for different networks in the plane or in the sphere can be carried out for networks in any surface. The constant value of $V - E + F$ for any network in a given surface is known as the *Euler characteristic* of that surface. (You have to make sure that the network genuinely 'covers' the entire surface, and is not just a plane network in a small portion of the surface.) In the case of the torus, the Euler characteristic works out to be 0. Hence this topo-

logical invariant serves to distinguish the torus from the sphere, whose Euler characteristic is 2.

Now we know three features to distinguish surfaces: number of edges, orientability, and the Euler characteristic. Are there any others? More to the point, do we *need* any others, or are these three enough to distinguish any pair of surfaces that are not topologically equivalent?

Perhaps surprisingly, the answer is that these three invariants are indeed all we need. Proving this fact was one of the great achievements of nineteenth-century mathematics.

The key to the proof was the discovery of what are known as *standard forms* for surfaces, particular

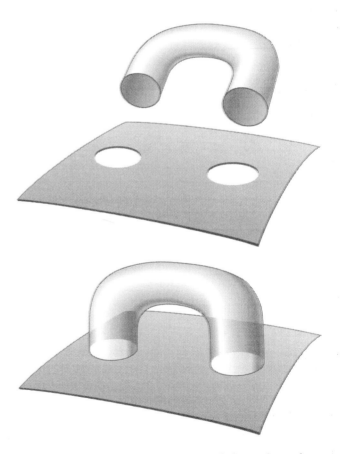

To attach a handle to a sphere, cut two holes in the surface and sew in a hollow, cylindrical tube to join them together.

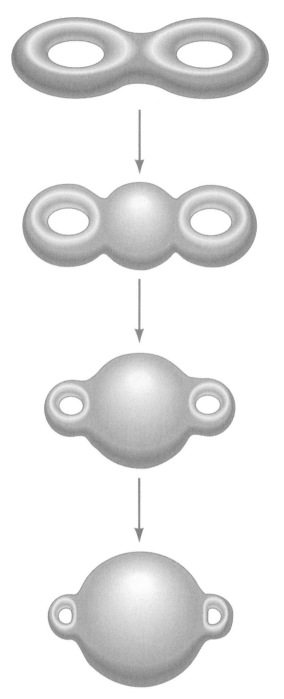

By blowing up the middle region and shrinking the two loops, a two-holed torus can be transformed into a sphere with two handles.

kinds of surfaces that suffice to characterize all surfaces. It was shown that any surface is topologically equivalent to a sphere having zero or more holes, zero or more 'handles', and zero or more 'crosscaps'. Thus, the topological study of surfaces reduces to an investigation of these modified spheres.

Suppose you have a standard surface that is topologically equivalent to a given surface. Then the holes in the standard surface correspond to the edges of the original surface. The simplest case has arisen already, when we were verifying Euler's formula for networks on a sphere. We removed one face, creating a hole, and then stretched out the edges of the hole to become the bounding edge of the resulting plane surface. Since this connection between holes in a sphere and edges of a surface is typical of the general case, from now on I shall restrict attention to surfaces having no edges, such as the sphere or the torus. Such surfaces are called *closed* surfaces.

To attach a *handle* to a surface, you cut two circular holes and sew on a cylindrical tube to join the two new edges, as shown in the figure on the previous page. Any closed orientable surface is topologically equivalent to a sphere with a specific number of handles. The number of handles is a topological invariant of the surface, called its genus. For each natural number $n \geq 0$, the standard (closed) orientable surface of genus n is a sphere to which are attached n handles. For example, the sphere is a standard orientable surface of genus 0; the torus is topologically equivalent to a sphere with one handle, the standard orientable surface of genus 1; and the two-holed torus is topologically equivalent to a sphere with two handles, the standard orientable surface of genus 2. The figure on this page illustrates a two-holed torus, made from some highly elastic material, that is manipulated to give a sphere with two handles.

The Euler characteristic of a sphere with n handles is $2 - 2n$. To prove this, you start with a (fairly large) network on a sphere (for which $V - E + F = 2$) and then add n handles, one by one. You take care to add each handle so that it connects two holes that result from the removal of two faces of the net-

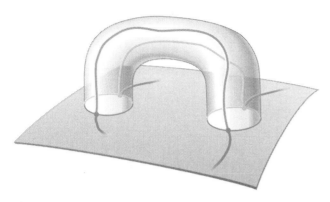

Adding handles to a sphere in order to calculate the Euler characteristic of the resulting surface.

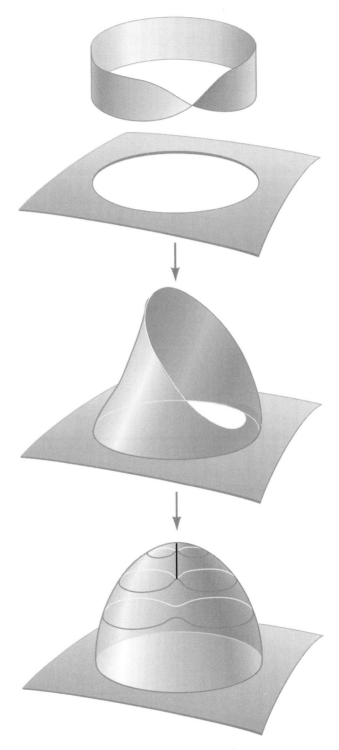

work. To ensure that the network on the resulting surface genuinely 'covers' the surface, you add two new edges along the handle, as shown in the figure above. Cutting the two holes initially decreases F by 2, sewing on the handle (with its new edges) increases both E and F by 2. The net effect of adding the handle, therefore, is that the quantity $V - E + F$ decreases by 2. This occurs every time you add a new handle; so, if n handles are added, the Euler characteristic decreases by $2n$, giving a final value of $2 - 2n$.

The standard nonorientable (closed) surface of genus n is a sphere to which n crosscaps have been added. To add a crosscap to a sphere, you cut a hole and sew on a Möbius band, as shown in the figure on the right. The entire edge of the Möbius band

To add a crosscap to a sphere, cut a hole in the surface and sew in a Möbius band. In three-dimensional space, the attachment of the Möbius band can only be achieved in a theoretical way, by allowing the band to intersect itself. In four dimensions, no such self-intersection is necessary. The crosscap is a surface that requires four dimensions to be constructed properly.

must be sewn to the circular hole; in three-dimensional space this can only be done if you allow the surface to intersect itself. To sew on the entire edge without such self-intersection you need to work in four-dimensional space. (Remember, any surface is two dimensional; the surrounding space is not part of the surface itself. Any surface other than the plane requires at least three dimensions to be constructed. The crosscap is probably the first surface you have encountered that needs four dimensions.)

In order to calculate the Euler characteristic for a sphere with one or more crosscaps, you start with a suitably large network in the sphere, and add the appropriate number of crosscaps, one at a time. Each crosscap replaces one complete face of the network: cut out the face, and sew in a Möbius band along the boundary of that face. Draw a new edge in the Möbius band, connecting the edge to itself, as shown directly below. Removing a face reduces F by 1; sewing on the Möbius band adds an edge and a face. Hence the net effect of adding a crosscap is to reduce $V - E + F$ by 1. Thus, the Euler characteristic of a sphere with n crosscaps is $2 - n$.

In particular, a sphere with a single crosscap has Euler characteristic $V - E + F = 1$, and a sphere

The Klein bottle. Popularly described as a container having no 'inside' or 'outside', this surface can be constructed in three-dimensional space only if you allow the surface to pass through itself. In four dimensions, it is possible to construct a Klein bottle without the need for such self-intersection. Topologically, you can obtain a Klein bottle by sewing together two Möbius bands along their (single) edges.

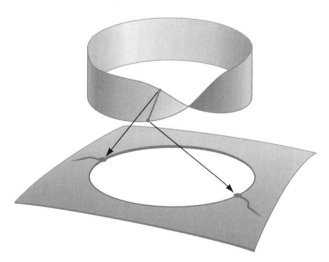

Adding crosscaps to a sphere in order to calculate the Euler characteristic of the resulting surface.

with two crosscaps has Euler characteristic 0. This last surface, which can also be constructed by sewing together two Möbius bands along their single edges, is popularly known as a Klein bottle, and is often depicted as in the photograph above. When regarded as some sort of vessel, it has neither 'inside' nor 'outside'. Again, the self-intersection is a result of trying to realize the surface in three-dimensional space; in four dimensions, there is no need for the surface to pass through itself.

In order to complete the classification of all surfaces, it is enough to demonstrate that any closed surface (i.e. any surface without edges) is topologically equivalent to one of the standard surfaces. The idea is to show that, given any closed surface, a process of successively removing pieces topologically equivalent to a cylinder or a Möbius band and replacing them by disks leads eventually to a sphere. A cylinder is replaced by two disks, a Möbius band by one. This process is known as *surgery*. Though the argument is not particularly difficult, there are many technical details, and I will not explain them here.

Manifolds

A surface can be regarded as made up of a number—possibly a large number—of small, virtually planar pieces sewn together. Within any one of the component pieces, the surface is just like a piece of the Euclidean plane. The global properties of the surface arise as a result of the manner in which all the component pieces are assembled together. For example, different means of assembly result in the distinction between a sphere and a torus. On any small region of either, the surface seems like the Euclidean plane, and yet globally the two surfaces are very different. We are familiar with this phenomenon from our own lives: based purely on our experiences within our everyday, local environments, there is no way we can tell whether the planet on which we live is planar, spherical, or shaped like a torus. This is why the notions and results of plane Euclidean geometry are so relevant to our everyday lives.

A distinction can be made between surfaces that are composed of pieces sewn together in a smooth fashion, without any sharp corners or folds, and surfaces assembled with sharp edges, such as polyhedra. Surfaces of the former kind are known as *smooth* surfaces. (Thus, mathematicians assign a technical meaning to the word 'smooth' in this context. Fortunately, this technical meaning accords with the everyday meaning.) In the case of a surface with a sharp edge at some join, the part of the surface surrounding that join does *not* resemble part of the Euclidean plane.

Based on the ideas discussed above, Riemann introduced the notion of a *manifold* as an important generalization of a surface to higher dimensions. A surface is a two-dimensional manifold, or 2-manifold for short. The sphere and the torus are examples of smooth 2-manifolds. An *n*-dimensional manifold, or *n*-manifold, consists of a number of small pieces glued together, each small piece being, to all intents and purposes, a small region of *n*-dimensional Euclidean space. If the 'seams' where the component pieces are joined together are free of 'sharp corners' or 'folds', the manifold is said to be smooth.

A fundamental question of physics is, what kind of 3-manifold is the physical universe in which we live? Locally, it looks like three-dimensional Euclidean space, as does any 3-manifold. But what is its *global* form? Is it everywhere like Euclidean 3-space? Or maybe it is a '3-sphere' or a '3-torus'? No one knows the answer.

The nature of the universe aside, the fundamental problem in manifold theory is to classify all possible manifolds. This means finding topological invariants that can distinguish between manifolds that are not topologically equivalent. Such invariants would be higher-dimensional analogues of orientability and the Euler characteristic, which served to classify all closed 2-manifolds. The classification problem is by no means solved. Indeed, mathematicians are still trying to overcome an obstacle encountered during the very first investigations into the task of classification.

Henri Poincaré was one of the first mathematicians to look for topological invariants applicable to higher-dimensional manifolds. In so doing, he helped to found the particular brand of topology now known as *algebraic topology*, which attempts to use concepts from algebra in order to classify and study manifolds.

One of Poincaré's inventions was the *fundamental group* of a manifold. The basic idea, illustrated

Henri Poincaré (1854–1912).

in the figure on page 190, is this. You fix some point, O, in the manifold, and consider all loops through the manifold that start and finish at O. Then you try to turn these loops into a group. That means you have to find an operation that can be used to combine any two of these loops into a third, and then verify that this operation satisfies the three axioms for a group. The operation that Poincaré considered is the 'group sum': if s and t are loops, the group sum $t + s$ is the loop that consists of s followed by t. This operation is associative, so you are already well on the way to having a group. Moreover, there is an obvious identity element: the 'null loop' that never even leaves the point O. So, if every element has an inverse, you will have your group. The idea then will be to see to what extent the algebraic properties of the fundamental group characterize the manifold.

The Four Color Theorem

A famous question in topology, formulated in 1852, asked for the minimum number of colors needed to draw a map, subject to the requirement that no two regions having a stretch of common border are colored the same way. Many simple maps cannot be colored using just three colors. On the other hand, for most maps, such as the county map of Great Britain shown here, four colors suffice. The *four color conjecture* said that four colors suffice to color *any* map in the plane. Over the years, a number of professional mathematicians attempted to prove the conjecture, as did many amateurs. One difficulty of the problem was that it asked about all possible maps, not just some particular maps. So there was no hope of proving that four colors suffice by looking at any particular map.

The problem is clearly a topological one. What counts is not the shape of the regions of the map but their *configuration*—which regions share a common border with which other regions. In particular, the number of colors required will not vary as you manipulate the surface on which the map is drawn—though the answer might vary from one kind of surface to another, say between maps drawn on a sphere and maps drawn on a torus. However, the minimum number of colors required to color any map is the same for maps drawn on the sphere as for maps drawn on the plane, so the four color conjecture may be stated equivalently for maps on the sphere.

In 1976, Kenneth Appel and Wolfgang Haken solved the problem, and the four color conjecture became the four color theorem. A revolutionary aspect of their proof was that it made essential use of a computer. The proof rested upon some previous work of Alfred Bray Kempe, a London barrister who, in 1879, had produced what turned out to be

There is an obvious candidate for the inverse of a given loop ℓ, namely the 'reverse loop', the loop that follows exactly the same path as ℓ but in the opposite direction. The reverse loop may reasonably be denoted by the symbol $-\ell$. The problem is, al-

arrangements, each requiring a detailed analysis. It required some 1,200 hours of computer time to carry out this analysis, but in the end the question was resolved. The published proof consisted of some 50 pages of text and diagrams, a further 85 pages containing 2,500 more diagrams, 400 microfiche pages giving details of various parts of the proof, plus the results of the computation, which had to be taken more or less at face value.

The problem of coloring maps extends naturally to maps drawn on nonplanar surfaces. At the turn of the century, Percy Heawood found a formula that, apart from one exception, seemed to give the minimum number of colors required to color any map on any given closed surface. For a closed surface having Euler characteristic n, the formula predicts the minimum number of colors to be

$$\frac{1}{2}(7 + \sqrt{49 - 24n}).$$

For example, according to this formula, the minimum number of colors required to color any map on a torus, for which $n = 0$, is 7. For the sphere, for which $n = 2$, the formula gives the answer 4. (Sadly for Heawood, he was not able to *prove* that his formula gave the right answer in the case of the sphere, so his formula did not help him to prove the four color conjecture.)

It is now known for certain that Heawood's formula gives the exact minimal number of colors in all cases except for the Klein bottle. For this surface, the Euler characteristic is 0, just as for the torus, so according to the formula, seven colors suffice; but any map on a Klein bottle can be colored using six colors.

a false proof of the theorem. Kempe's main idea was this: first you show that in any map that required five colors to color properly, certain special configurations of countries must occur. Then, by examining each of these special configurations in turn, you show that none of them can occur in a map that requires five colors. Taken together, these two results imply that no map requires five colors, and the four color theorem is proved.

Kempe's overall strategy was correct, but unfortunately it turned out that the class of special configurations involved contains some 1,500

though $-\ell$ 'undoes' the effect of ℓ, the combination $-\ell + \ell$ is not the null path, just as flying from New York to San Francisco and then flying back again is not the same as never leaving New York in the first place. True, in both cases you start and fin-

ish in New York; but what happens in between is quite different.

The way out of this dilemma is to declare any two loops to be identical if one may be continuously deformed into the other within the manifold. For

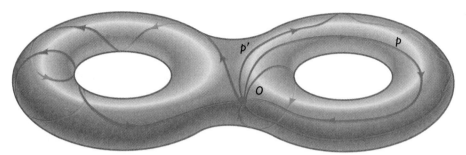

The fundamental group of a manifold.

example, in the figure on this page, the path p may be continuously deformed into p' within the manifold. This does the trick, since $-\ell + \ell$ can clearly be transformed continuously to the null loop.

A continuous transformation from one loop or path to another is known as a *homotopy*, and Poincaré's fundamental group, obtained in this manner, is known as a *homotopy group* of the manifold. In the highly simplified case where the manifold is a circle (a 1-manifold), the only difference between two loops is the number of times each winds around the circle, and the fundamental group in this case turns out to be the group of the integers under addition.

On their own, the fundamental groups do not suffice to classify manifolds. But the general idea was a good one, so Poincaré and other mathematicians took it further. By using n-dimensional spheres instead of 1-dimensional loops, for each natural number n, they constructed what is known as a *higher homotopy group* of dimension n. Any two topologically equivalent manifolds must have the same homotopy groups, and the question was, does the collection of *all* homotopy groups suffice to distinguish any two manifolds that are not topologically equivalent?

With the classification problem for 2-manifolds already solved, the first case to look at was in dimension 3. A special instance of that case was the question, if a 3-manifold $\mathcal{M}$ has the same homotopy groups as the 3-sphere, $\mathcal{S}^3$, is $\mathcal{M}$ topologically equivalent to $\mathcal{S}^3$? It was Poincaré himself who first raised this question, in 1904, and the conjec-

ture that the answer is yes became known as the Poincaré conjecture.

The conjecture generalizes to the case of n-manifolds in a straightforward manner: if an n-manifold $\mathcal{M}$ has the same homotopy groups as the n-sphere, $\mathcal{S}^n$, is $\mathcal{M}$ topologically equivalent to $\mathcal{S}^n$?

Using the classification of 2-manifolds, the answer may be shown to be yes in the case $n = 2$. (The question reduces to looking at homotopy groups associated with the standard surfaces.) But for many years, no one was able to make much headway with any of the higher-dimensional cases, and the Poincaré conjecture began to achieve the same status in topology as Fermat's last theorem in number theory. In fact, this comparison does not really do full justice to the Poincaré conjecture. Whereas Fermat's last theorem grew more famous the longer it remained unproved, it did not have any major consequences. The Poincaré conjecture, in contrast, is the key to a whole new area of mathematics, a fundamental obstacle that stands in the way of further progress in our understanding of manifolds.

Given the relative ease with which the 2-manifold case of the conjecture can be proved, one might imagine that the next case to be solved would be dimension 3, then dimension 4, and so forth. But issues of dimension do not always work that way. Though the complexity and difficulty of problems generally does increase as you go up through dimensions 1, 2, and 3, it can happen that things simplify enormously when you get to dimension 4 or 5 or so. The additional dimensions seem to give you more room to move about, more scope to develop tools to solve your problem.

This is exactly what happened in the case of the Poincaré conjecture. In 1961, Stephen Smale proved the conjecture for all dimensions n from $n = 7$ upward. Soon afterward, John Stallings pushed the result down to include the case $n = 6$, and Christopher Zeeman took it a step further to include $n = 5$. Only two more cases were left to be proved!

A year went by, then two, then five. Then a decade. Then two decades. Further progress began to seem hopeless.

Then, in 1982, Michael Freedman finally broke the deadlock, finding a way to establish the truth of the Poincaré conjecture for dimension 4. That left only the case $n = 3$. And there, to everyone's frustration, the matter remains to this day. With all cases of the Poincaré conjecture verified apart from dimension 3, it is tempting to assume that the conjecture is indeed true for all dimensions. And it is probably true that most topologists expect this to be the case. But expectation is not proof, and for the moment the Poincaré conjecture remains one of the greatest, if not *the* greatest, unsolved problems in topology.

One possible way to approach the problem, and indeed the overall classification for 3-manifolds, is to make use of techniques from geometry. This approach is reminiscent of Klein's Erlangen program, in which group theory was used to study geometry—see page 152. This, at least, was the proposal put forward by the mathematician William Thurston during the 1970s. Despite the fact that topological properties are highly *non*geometric, Thurston thought that geometric patterns might nevertheless prove useful for the study of 3-manifolds.

Such a proposal is not easily followed. For one thing, in three dimensions there are eight different geometries to deal with, as Thurston himself proved in 1983. Three of these correspond to the three planar geometries, namely Euclidean 3-space, elliptic 3-space (which corresponds to two-dimensional Riemannian geometry), and hyperbolic 3-space (which corresponds to two-dimensional hyperbolic geome-

try). The remaining five geometries are new ones, which arose as a result of Thurston's investigation.

Though Thurston's program is by no means complete, considerable progress has been made, demonstrating once again the incredible power of cross-fertilization in mathematics, when patterns in one area are applied to another. In this case, Thurston analyzes the group-theoretic patterns of possible geometries (remember, any geometry is essentially determined by a particular group of transformations) and then applies those geometric patterns to the topological study of 3-manifolds.

It should be said that the expectation that the one remaining case of the Poincaré conjecture will turn out to be true is certainly not based on the fact

William Thurston of the University of California at Berkeley.

that all the other cases have been proved. If ever topologists had thought that all dimensions behaved in more or less the same way, they were forced to change their views radically by an unexpected, and dramatic, discovery made in 1983, by a young English mathematician named Simon Donaldson.

Physicists and engineers make frequent use of the differential calculus in the study of Euclidean space of dimension 3 or more. Generalizing the notion of differentiation from the two-dimensional case described in Chapter 3 is relatively straightforward. However, physicists, in particular, need to be able to use techniques of the differential calculus on smooth manifolds other than Euclidean n-space. Since any n-manifold can be split up into small pieces that each look like Euclidean n-space, for which we know how to do differentiation, it is possible to apply the differential calculus to a manifold in a strictly local way. The question is, can differentiation be carried out globally? A global scheme for differentiation is called a *differentiation structure.*

For instance, in the case of the regular 2-sphere, lines of latitude and longitude drawn on the sphere provide a coordinate system for any small portion of the surface that supports differentiation in the usual way. Because the coordinate system is uniform, the entire latitude-longitude system in effect provides a global *differentiation structure,* which is illustrated in the picture on page 172.

In fact, drawing lines of latitude and longitude is essentially the only way to obtain a differentiation structure on the 2-sphere. Indeed, by the mid-1950s it was known that any smooth 2- or 3-manifold could be given a unique differentiation structure, and it was assumed that this result would eventually be extended to higher dimensions. But, to everyone's surprise, in 1956 John Milnor discovered that the 7-sphere could be given 28 distinct differentiation structures, and soon afterward similar results were discovered for spheres of other dimensions.

Still, topologists could comfort themselves that these new results did not apply to Euclidean n-space itself. It was surely the case, they thought, that there

was only one differentiation structure on these familiar spaces, the spaces for which the original methods of Newton and Leibniz apply.

Or was it? Certainly, it was known that both the Euclidean plane and Euclidean 3-space have a unique differentiation structure, namely the standard one. It was also known that the standard differentiation structure was unique for Euclidean n-space for all values of n other than 4. Curiously, however, no one had been able to give a proof for the four-dimensional case. But it was surely just a matter of time before someone hit upon the right combination of ideas, wasn't it? The mathematicians were all the more frustrated at their seeming inability to provide an answer because this was the very space of most concern to physicists working in four-dimensional space-time. The physicists were waiting for their colleagues the mathematicians to resolve the matter.

But this turned out to be a rare occasion when the usual way of doing business was reversed. Instead of the physicists making use of new ideas from mathematics, methods of physics came to the rescue of the mathematicians. In 1983, using ideas from physics known as Yang–Mills gauge fields, which had been introduced in order to study the quantum behavior of elementary particles, together with the methods Michael Freedman had developed in order to prove the Poincaré conjecture in dimension 4, Simon Donaldson showed how to construct a *nonstandard* differentiation structure on Euclidean 4-space, in addition to the usual one.

In fact, the situation rapidly became even more bizarre. Subsequent work by Clifford Taubes demonstrated that the usual differentiation structure on Euclidean 4-space is just one of an infinite family of different differentiation structures! The results of Donaldson and Taubes were completely unexpected, and quite contrary to everyone's intuitions. It seems that, not only is Euclidean 4-space of particular interest because it is the one in which we live (if you include time), it is also the most interesting—and the most challenging—for the mathematician.

Knots

The first book on topology was *Vorstudien zur Topologie*, written by Gauss' student Listing, and published in 1847. A considerable part of this monograph was devoted to the study of knots.

To the layperson, the first figure on this page illustrates two typical knots, the familiar overhand knot and the figure-of-eight. Most people would agree that these are *different* knots. But what exactly does it mean for two knots to be different? Not that they are made from different pieces of string, or whatever. Nor does the actual *shape* of the string matter. If you were to tighten either of these knots, or change the size or shape of the loops, then the overall appearance would change, perhaps dramatically, but it would still be the same knot. No amount of tightening or loosening or rearranging would seem to turn an overhand knot into a figure-of-eight knot. The distinctive feature of a knot, surely, is its 'knottedness', the manner in which it loops around itself. It is this abstract pattern that mathematicians set out to study when they do 'knot theory'.

Since the knottedness of a string does not change when you tighten it, loosen it, or manipulate the shape of the individual loops, knot patterns are topological. You can expect to make use of some of the ideas and methods of topology in studying knots. But you have to be careful. For one thing, there is surely a very easy way to manipulate, topologically, the overhand knot so that it becomes a figure-of-eight knot: simply untie the overhand knot and retie it as a figure-of-eight. There is no cutting or tearing involved in this process; topologically speaking, the procedure is perfectly in order. But, obviously, if you want to make a mathematical study of knots, you want to exclude the case of transforming one knot into another by untying the first and then retying it in the form of the second.

So, when mathematicians study knots, they demand that the knot has no free ends, as shown in the second figure on this page. The two knots illustrated result from attaching together the free ends of the overhand and figure-of-eight; they are called

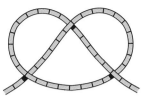

The overhand

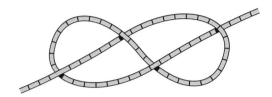

The figure-of-eight

Two everyday knots: (i) the overhand, and (ii) the figure-of-eight.

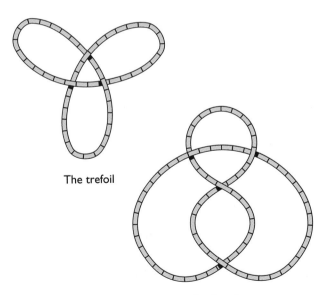

The trefoil

The four-knot

Two mathematical knots: (i) the trefoil, and (ii) the four-knot. Mathematical knots consist of closed loops in space, and knot theorists study the patterns of 'knottedness' exhibited by different knots.

the trefoil and the four-knot, respectively. In the case of physical knots, made out of string, attaching the two ends would mean gluing them together.

Restricting attention to knots tied in *closed loops* of string solves the problem of trivializing the study by being able to untie and retie, and yet it clearly retains the essential notion of knottedness. It is surely impossible to transform the trefoil knot shown on the preceding page into the four-knot beneath it. (Try it for yourself and see. Construct a trefoil out of string, join the ends together, and then attempt to turn your knot into a four-knot, without undoing the two ends.)

Having decided to ignore the material a knot is constructed from, and to insist that a knot has no free ends, you arrive at the mathematician's definition of a knot: it is a closed loop in three-dimensional space. (From this standpoint, the two 'knots' shown at the top of the preceding page are not knots at all.) As a 'loop' in space, a mathematical knot will have no thickness, of course; it will be a one-dimensional entity, a 1-manifold to be precise.

A computer-generated image of the trefoil knot.

The task now is to study the *patterns* of knots. This means ignoring issues of tightness, size, the shape of individual loops, and the position and orientation of the knot in space; mathematicians do not distinguish between knots that are topologically equivalent.

But just what is meant by that last phrase, 'topologically equivalent'? There is certainly a very simple way to transform, say, a trefoil into a four-knot in a topologically legitimate fashion. Cut the string, untie it, retie it in the form of a four-knot, and then fasten the free ends back together again. Points close together before this process remain close together afterward, so this is a permissible topological tranformation. But it clearly violates the spirit of what we are trying to do: the one thing we do not want to allow in the study of knots is cutting.

The point about a mathematical knot is that its pattern arises from the manner in which it is situated in the surrounding three-dimensional space. That pattern is topological, but the topological transformations that are relevant are transformations of all of 3-space, not just the knot. When mathematicians speak of two knots being topologically equivalent (and hence, in fact, being the 'same' knot), they mean that there is a topological transformation of 3-space that transforms one knot into the other.

Though this official definition of 'knot equivalence' is important in the detailed, mathematical study of knots, it is not very intuitive. But the essence of the definition is that it excludes cutting the knot loop, but allows any other topological manipulation. Indeed, it is only in more advanced work in knot theory that topologists look closely at the 'whole of space', as they examine the complicated 3-manifold that remains when a knot is *removed* from 3-space.

The study of knots is a classic example of the way mathematicians approach a new area of study. First, a certain phenomenon is observed, in this case knottedness. Then, the mathematician abstracts away from all those issues that appear irrelevant to the study, and formulates precise definitions of the crucial notions. In this case, those crucial notions are knots and knot-equivalence. The next step is to find ways to describe and analyze the different kinds of knot—the different knot patterns. In particular, knot theorists seek ways to classify all knots in terms of *knot invariants*.

For example, what distinguishes the trefoil knot from the *null knot*—the unknotted loop? Of course, they *look* different. But, as mentioned earlier, what a knot looks like—that is to say, the manner in which a particular knot is laid out, or *presented*—is not important. The question is, can the trefoil be manipulated into an unknotted loop without cutting the loop? It certainly seems as though it cannot. Moreover, if you were to make a trefoil out of string and play around with it for a while, you might find you were unable to 'unknot' it. But this does not amount to a *proof*. Maybe you simply have not tried the right combination of moves.

(Actually, it is arguable that, in the case of a very simple example such as the trefoil, mental or physical manipulation does amount to a proof, in all but the strictest sense of formal logic—a standard of proof that almost no real theorem of mathematics ever adheres to. But in the case of more complicated knots, such an approach would not constitute a proof. Besides, what is wanted is not a method for coping with one particular, very simple example, but general methods that work for all knots, including those nobody has seen yet. One should always be cautious in using examples. To serve as such, an example has to be simple; but the aim is to help understand the underlying issues, issues that apply to more complicated cases.)

A more reliable method to distinguish two knots would be to find some knot invariant for which they differ, a knot invariant being any property of a knot that does not change when you subject the knot to any permissible manipulation. In order to look for knot invariants, you must first find some way to represent knots. An algebraic notation might help eventually, but at the outset of the study, the most obvious representation is a diagram. Indeed, I have already presented two *knot diagrams* on

page 193. The only modification that mathematicians make when they draw such diagrams is that they do not try to draw a picture of a physical knot, made from string or rope, but give a simple line drawing that indicates the knot pattern itself. The figure on this page gives some examples, including the mathematician's version of the trefoil, illustrated

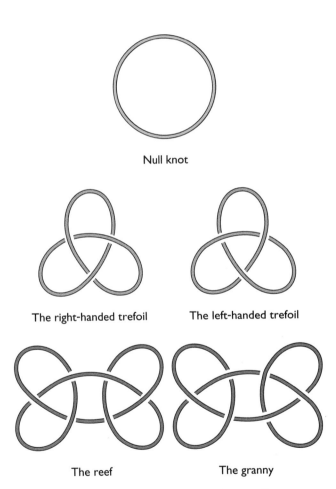

Null knot

The right-handed trefoil   The left-handed trefoil

The reef   The granny

Five simple knots, as a knot theorist would typically represent them. After a few moments consideration, it seems obvious—but it is not easy to prove—that each of these five knots is different from all the others; that is to say, no amount of manipulation of any one of these knots will transform it into one of the others.

previously on page 193. The lines are broken to indicate where the knot passes over itself. Such a diagrammatic representation of the knot is often referred to as a *presentation* of the knot.

One way to try to understand the structure of a complicated knot is to try to split it up into a number of smaller, simpler knots. For example, both the reef knot and the granny knot can be 'split up' into two trefoils. Expressing this the other way round, one way to construct a (mathematical) reef or a granny is to tie two trefoils on the same 'mathematical string', and then join the free ends. It is natural to describe this process of tying two knots onto the same string as constructing the *sum* of the two knots.

This summation operation is associative, and the null knot is obviously an identity operation. At which point, the mathematician, ever on the lookout for new patterns, will begin to wonder if this is yet another example of a group. All that would be required is for every knot to have an 'inverse'. Given any knot, is it possible to introduce another knot on the same mathematical string so that the result is the null knot? If it were, then by manipulating the string, the knotted loop could be transformed into an unknotted loop. Though stage magicians know how to tie certain kinds of 'knot' with this property, it is not the case that every knot has an inverse, and, in fact, the magician's 'knotted string' is not knotted at all, it just looks as though it is. The operation of summing two knots does not give rise to a group.

But just because one avenue does not lead anywhere, it does not mean you should give up looking for familiar patterns. Knot summation may not produce the group pattern, but there are other algebraic patterns that might arise. Having seen that reef and granny knots can be split up into sums of simpler knots, you might consider introducing the notion of a *prime knot*, a knot that cannot be expressed as the sum of two simpler knots.

Before you can proceed in that direction, you ought to say just what you mean by 'simpler' in this context. After all, it is easy to take an unknotted

loop and manipulate it into what looks like a fiendishly complex knot—fine necklace chains have a habit of achieving such a state without any apparent help from a human hand.

To define the complexity of a knot, mathematicians associate with any knot a positive whole number, called the crossing number of the knot. The procedure is as follows. If you take a knot diagram, you can add up the total number of crossings, points where the line passes over itself. (The same number can also be described as the number of breaks in the line, as drawn. You have to first manipulate the knot diagram so that you never have three lines crossing at the same point.) The number of crossings provides you with a measure of the complexity *of the diagram*. Unfortunately, it tells you little about the actual knot. The problem is, the same knot can have infinitely many different numbers associated with it in this way: you can always increase the number by 1 as many times as you like, without changing the knot, by the simple act of introducing new twists in the loop.

But, for any knot, there will be a unique, *least* number you can obtain in this way. *That* number clearly is a measure of the knot's complexity; it is the number of crossings in a diagram that represents the knot in the 'simplest' possible way, devoid of any superfluous twists in the loop. It is this least number of crossings that is called the crossing number. It tells you how many times the loop is *forced* to cross itself in order to produce the knot, regardless of how many times it actually does cross itself in a particular presentation of the knot. For example, the crossing number of the trefoil is 3, and that of both the reef and the granny is 6.

You now have a way to compare two knots: knot A is 'simpler' than knot B if A has smaller crossing number than B. You can proceed to define a prime knot as one that cannot be expressed as the sum of two simpler knots (neither is the null knot).

A great deal of the early work in knot theory consisted of an attempt to identify all the prime knots having a given crossing number. By the turn of the century, many prime knots had been identified having crossing numbers up to 10, with the results being presented as tables of knot diagrams.

The work was fiendishly difficult. For one thing, in all but the simplest cases, it is extremely hard to tell if two different-looking diagrams represent the same knot, so no one was ever sure if the latest table had duplicate entries. Still, with work of J. W. Alexander and G. B. Briggs in 1927, mathematicians knew there were no duplications in the tables for crossing numbers as far as 8, and, a short while later, H. Reidemeister sewed up the case for crossing number 9 as well. The case of crossing number 10 was finally settled in 1974, by K. A. Perko. All of this progress was achieved by using various knot invariants to distinguish between different knots.

The crossing number is a very crude knot invariant. Though knowledge that two knots have different crossing numbers does tell you that the knots are definitely not equivalent, and, indeed, provides a way to compare the complexities of the knots, there are simply too many knots having the same crossing number for this means of knot classification to be of much use. For instance, there are 165 prime knots with crossing number 10.

One reason why the crossing number is such a weak invariant is that it simply *counts* crossings, it does not attempt to capture the pattern of crossings as the knot weaves around itself. One way to overcome this deficiency was discovered by J. W. Alexander in 1928. Starting with the knot diagram, Alexander showed how to calculate not a number but an algebraic polynomial, the *Alexander polynomial*. The exact details of how to do this are not important in the present context; for the trefoil, the Alexander polynomial is

$$x^2 - x + 1,$$

and for the four-knot it is

$$x^2 - 3x + 1.$$

When two knots are added together, multiplying the two Alexander polynomials produces the Alexander polynomial of the sum knot. For example, the Alexander polynomial of both the reef and the granny knot, each of which is a sum of two trefoils, is

$$(x^2 - x + 1)^2 = x^4 - 2x^3 + 3x^2 - 2x + 1.$$

In capturing, in an algebraic fashion, something of the way in which a knot winds around itself, the Alexander polynomial is certainly a useful knot invariant. And it is fascinating to know that a knot pattern can be partially captured by an algebraic pattern. But the Alexander polynomial is still somewhat crude; it does not capture enough of the pattern of a knot to distinguish, say, a reef knot from a granny, something done with ease by any child who has been camping.

Other attempts to find simple knot invariants also tended to fall short when it came to distinguishing between the reef knot and the granny, but for all their shortcomings, the various approaches highlight yet again the manner in which the different patterns the mathematician weaves find application in many distinct areas.

The Alexander polynomials themselves are derived from another, fairly powerful knot invariant called the *knot group*. This is the fundamental group (or homotopy group) of the knot *complement*, the 3-manifold that is left when the knot itself is removed. The members of this group are closed, directed loops that start and finish at some fixed point not on the knot, and which wind around the knot. Two loops are declared identical in the knot group if one can be transformed into the other by manipulating it in a manner that does not involve cutting or passing through the knot. The figure on the next page illustrates the knot group for the trefoil.

The knot group provides the mathematician with a means of classifying knots according to properties of groups. An algebraic description of the knot group can be derived from the knot diagram.

Another ingenious way to classify knots is to construct, for a given knot, an orientable (i.e. 'two-sided') surface having the knot as its only edge, and

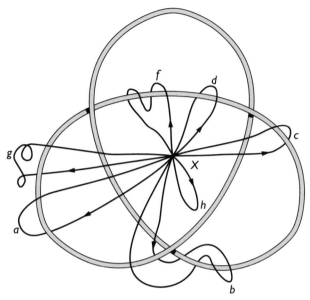

The knot group for the trefoil. The members of the group are closed, directed loops that start and finish at the point X. Loops a, b, and g are regarded as the same, since one can be transformed to the other without cutting or passing through the knot. Loops c and d are regarded as distinct, since they pass around the knot in opposite directions. Loop h is identified with the null loop of zero length, the identity element of the group. The group operation is the combination of loops, with the 'sum' x + y of loops x and y consisting of loop y followed by loop x. (The intermediate step of passing through the base-point X is ignored when two loops are combined.) For example, d + d = f and c + d = h.

all of that changed in 1984, when a New Zealand mathematician by the name of Vaughan Jones discovered a quite new class of polynomial invariant for knots.

The discovery was made quite by chance. Jones was working on a problem in analysis that had applications in physics. The problem concerned mathematical structures known as von Neumann algebras. In looking at the way these von Neumann algebras were built up from simpler structures, he discovered patterns that colleagues found reminiscent of some patterns to do with knots, discovered by Emil Artin in the 1920s. Sensing that he had stumbled onto an unexpected, hidden connection, Jones consulted knot theorist Joan Birman, and the rest, as they say, is history. Like the Alexander polynomial, the Jones polynomial can be obtained from the knot diagram. But, far from being a simple variant of the Alexander polynomial, as Jones himself had at first thought, his polynomial was something quite new.

In particular, the Jones polynomial can distinguish between a reef knot and a granny. The difference between these two knots depends on the orientation of the two trefoils relative to each other. As illustrated in the figure on page 195, a trefoil knot can wrap around in one of two ways, resulting in two trefoils, each the mirror image of the other. The Alexander polynomial does not distinguish between these two varieties of trefoil, and hence cannot distinguish between a reef knot and a granny knot. The Jones polynomial does distinguish between the two trefoils, however. The two Jones polynomials are:

$$x + x^3 - x^4,$$

$$x^{-1} + x^{-3} - x^{-4} = \frac{1}{x} + \frac{1}{x^3} - \frac{1}{x^4}.$$

Strictly speaking, the second of these is not a polynomial, since it contains negative powers of the variable, x. But in this case, mathematicians use the word anyway.

to take the genus of that surface as an invariant of the knot. Since there may be more than one surface that can be associated with the same knot in this fashion, you take the smallest genus number that can arise this way. The resulting number, called the 'genus of the knot', is a knot invariant.

None of the knot invariants mentioned so far can distinguish a reef knot from a granny, and for many years there seemed to be no hope of finding a simple way to do this. (Knot theorists could make the distinction using more complex invariants.) But

Genetic Knots

A single strand of human DNA can be as long as one meter. Coiled up, it can fit into a cell nucleus having a diameter of about 5 millionths of a meter. Clearly, the DNA molecule has to be pretty tightly interwoven. And yet, when the DNA divides to give two identical copies of itself, these two copies slide apart in an effortless way. What kind of knotting permits this smooth separation to happen? This is just one of many questions that face biologists in their quest to understand the secrets of life itself.

It is a question that might be answered with help from mathematics. Since the mid-1980s, biologists have teamed up with knot theorists in an attempt to understand the knot patterns nature uses in order to store genes. By isolating single DNA strands, fusing their ends to create a 'mathematical' knot, and then examining them under a micro-

scope, it has been possible to apply mathematical methods, including the Jones polynomials, in order to classify and analyze these fundamental patterns. The drawing on the right shows the knot structure exhibited by the strand of DNA in the electron microscope photo on the left.

In fact, not only was Jones' initial breakthrough significant in itself, it opened the way to a whole array of new polynomial invariants, and led to a dramatic rise in research in knot theory, some of it spurred on by the growing awareness of exciting new applications in both biology (see the box on this page) and physics (of which more in a moment).

If all of this was not enough, in 1987 further polynomial invariants for knots were found that were based on ideas from statistical mechanics, an area of applied mathematics that studies the molecular behavior of liquids and gases. And, not long afterward it was observed that the knot-theoretic pattern captured by the Jones polynomial itself arises in statistical mechanics. Knots, it seems, are everywhere. More precisely, the *patterns* that knots exhibit are everywhere.

The ubiquity of knots was illustrated in a particularly dramatic and far-reaching fashion by the rapid rise of *topological quantum field theory*, a new theory of physics developed by Edward Witten in the late 1980s. The mathematical physicist Sir Michael Atiyah was the first to suggest that the mathematical patterns captured by the Jones polynomial might be useful in trying to understand the structure of the physical universe. In response to Atiyah's suggestion, Witten came up with a single, very deep theory that generalizes, and builds upon, the patterns captured by quantum theory, the Jones polynomials, and the fundamental work of Simon Donaldson mentioned in the last section. This powerful new synthesis of ideas has provided physicists with a completely new way to regard the universe we live in, and has helped mathematicians obtain

Knot Theories of the Universe

Readers who were amused by Plato's and Kepler's theories of matter, described on pages 113 and 114, should consider some more recent theories.

In 1867, Lord Kelvin put forward an atomic theory proposing that atoms were knots in the ether. Known as the theory of vortex atoms, there were some sensible reasons for this proposal. It explained the stability of matter, and it provided a large collection of different atoms, taken from the rich collection of knots in the process of being classified by knot theorists at that time. It also provided an explanation of atomic vibrations, manifested in spectral lines. Kelvin's theory was taken sufficiently seriously to provide the impetus for some of the early mathematical work on the classification of knots; in particular, his collaborator P. G. Tait produced extensive tables of different knots. But for all its mathematical elegance, the theory of vortex atoms went the same way as Plato's atomic theory. It was eventually replaced by the idea put forward by Niels Bohr, of the atom as a miniature solar system.

These days, with Bohr's theory having been abandoned as too naive, knot theory has once again come to the fore. Physicists now suggest that matter is made up of 'superstrings', tiny, knotted, closed loops in space-time, whose properties are closely bound up with their degree of knottedness.

new insights into the theory of knots. The result has been a fruitful merger of topology, geometry, and physics that promises to lead to many further discoveries in all three disciplines.

In developing the mathematical theory of knots, mathematicians have thus created new ways of understanding certain aspects of the world, both the living world of DNA and the physical universe we live in. After all, what is understanding other than a recognition of a pattern of some kind or another?

Fermat's Last Theorem Again

Now, at last, it is possible to complete the account of Fermat's last theorem, commenced in Chapter 1.

The problem Fermat left behind, you will recall, is to prove that if n is greater than 2, the equation

$$x^n + y^n = z^n$$

has no (nontrivial) whole-number solutions. Because of our everyday familiarity with the whole numbers, the simplicity of the question might suggest that it should not be too difficult to find a proof. But this impression is illusory. The problem with many such ad hoc questions is that, in order to find an answer, you have to unearth deep and hidden patterns. In the case of Fermat's last theorem, the relevant patterns proved to be many and varied, and very deep indeed. In fact, it is doubtful if more than a few dozen mathematicians in the entire world are in a position to fully understand much of the recent work on the problem. What makes giving a brief sketch of this work worthwhile is that it provides a powerful illustration of the way that apparently different areas of mathematics have deep, underlying connections.

The starting point for most of the past fifty years' work on the problem is to recast Fermat's claim as one about rational-number solutions to equations. First, notice that finding *whole-number* solutions to an equation of the form

$$x^n + y^n = z^n$$

(including the case $n = 2$) is equivalent to finding *rational-number* solutions to the equation

$$x^n + y^n = 1.$$

For, if you have a whole-number solution to the first equation, say $x = a$, $y = b$, $z = c$, where a, b, c are whole numbers, then $x = a/c$, $y = b/c$ is a rational-number solution to the second equation. For example,

$$3^2 + 4^2 = 5^2,$$

so $x = 3$, $y = 4$, $z = 5$ is a whole-number solution to the first equation. Dividing through by the solution value for z, namely 5, you get a rational-number solution to the second equation, namely $x = \frac{3}{5}$, $y = \frac{4}{5}$:

$$\left(\frac{3}{5}\right)^2 + \left(\frac{4}{5}\right)^2 = 1.$$

And if you have a rational-number solution to the second equation, say $x = a/c$, $y = b/d$, so that

$$\left(\frac{a}{c}\right)^n + \left(\frac{b}{d}\right)^n = 1,$$

then by multiplying the two solution numbers by the product of their denominators, cd (actually, the least common multiple will do), you get a whole-number solution to the first equation, namely $x = ad$, $y = bc$, $z = cd$:

$$(ad)^n + (bc)^n = (cd)^n.$$

In fact, it was the problem of finding *rational-number* solutions to equations that led to Fermat's last theorem in the first place. You may recall that a note scribbled in Fermat's copy of Diophantus' book *Arithmetic* started the whole affair. A great deal of that book concerned methods for finding rational-number solutions to equations. (In Diophantus' day, there would have been no need to stipulate 'rational number' here, since at that time it was still assumed that all numbers were rational.) Over the years, it had become commonplace to pose questions in terms of finding *whole*-number solutions of equations rather than rational solutions, and such questions are nowadays referred to as Diophantine problems. In the case of the problem that led to Fermat's last theorem, the two formulations are equivalent, as indicated above.

Unfortunately, though posing the question in terms of whole numbers might make the problem *look* simpler, especially to people who are uneasy dealing with fractions, for Fermat's last theorem this turned out to be exactly the wrong choice.

By formulating Fermat's problem as one about rational solutions, patterns of geometry and topology may be brought to bear. For instance, the equation $x^2 + y^2 = 1$ is the equation of the circle of radius 1, having its center at the origin. Asking for rational-number solutions to this equation is equivalent to asking for points on the circle whose coordinates are both rational. Because the circle is such a special mathematical object—to many people the most 'perfect' of all geometric objects—it turns out to be an easy task to find such points, using the following simple, geometric pattern.

Start off by choosing some point P on the circle. Any point will do. In the figure on this page, P is the point $(-1, 0)$, since choosing P to be this point makes the problem a bit simpler. The aim is to find points on the circle whose coordinates are

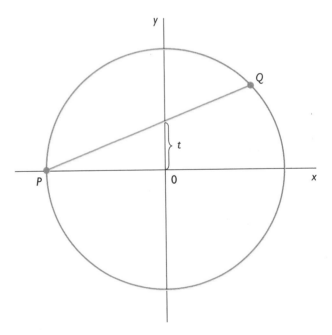

A geometric method to determine Pythagorean triples.

both rational. For any point Q on the circle, draw the line from P to Q. This line will cross the y-axis at some point. Let t denote the height of this crossing point above or below the origin. It is then an easy exercise in algebra and geometry to verify that the point Q will have rational coordinates if, and only if, the number t is rational. So, in order to find rational solutions to the original equation, all you need to do is draw lines from the point P to cross the y-axis a rational distance t above or below the origin, and the point Q where your line meets the circle will have rational coordinates. Thus you have a rational solution to the equation.

For instance, if you take $t = \frac{1}{2}$, a little computation shows that the point Q has coordinates $(\frac{3}{5}, \frac{4}{5})$. Similarly, $t = \frac{2}{3}$ leads to the point $(\frac{5}{13}, \frac{12}{13})$, and $t = \frac{1}{6}$ gives $(\frac{35}{37}, \frac{12}{37})$. These correspond to the (whole-number) Pythagorean triples $(3, 4, 5)$, $(5, 12, 13)$, and $(35, 12, 37)$, respectively. In fact, if you analyze this geometric approach, you will see that it leads to the formula for generating all the Pythagorean triples, given on page 30.

In the special case of exponent $n = 2$, therefore, the nice properties of the circle allow you to investigate the rational solutions to the equation

$$x^n + y^n = 1$$

by means of geometry. But there is no such easy analysis for all other values of n, for which the curve is by no means the simple, elegant circle. Recasting the problem in geometric terms, as looking for points on the curve

$$x^n + y^n = 1$$

that have rational coordinates, is still the right direction in which to proceed. But when n is bigger than 2, this step is just the start of a long and tortuous path.

The problem is that, lacking the nice geometric structure of the circle, the curves that you obtain seem no easier to analyze than the original equation. Faced with this hurdle, most mortals would

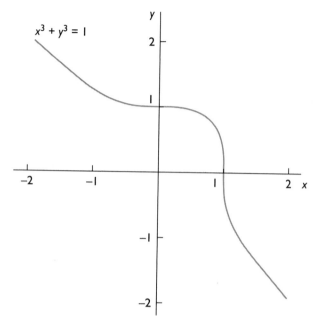

A Fermat curve: the curve $x^3 + y^3 = 1$.

give up and try something else. But if you were one of the many mathematicians whose work has contributed in a significant way to our present understanding of Fermat's last theorem, you would not. You would press forward, looking for additional structure other than the geometry of the curve. Your hope would be that the increased complexity of the additional structure would provide useful patterns that help your overall understanding, and eventually lead to a proof.

First of all, you can generalize the problem by allowing *any* polynomial in two unknowns. For any such equation, you can ask whether there are any rational solutions, and maybe by looking at the class of *all* equations, you will be able to discern patterns that will enable you to solve Fermat's original problem. It turned out, however, that this degree of generalization is not sufficient: you need quite a lot more structure before useful patterns start to appear. Curves just do not seem to exhibit enough useful patterns.

So, as a further generalization, suppose that the unknowns x and y in the equation are regarded as ranging not over real numbers but over complex numbers. Then, instead of giving rise to a *curve*, the equation will determine a *surface*, a closed, orientable surface to be precise. Examples of two of the surfaces that arise in this way are illustrated below. (Actually, not all equations give rise to a nice, smooth surface, but with some extra effort it is possible to patch things up so that everything works out.) The crucial point about this step is that surfaces are nice, intuitive objects that exhibit many useful patterns, and for which there is a wealth of mathematical theory available to be used.

For instance, there is the well-worked-out classification theory for surfaces: every closed, orientable smooth surface is topologically equivalent to a sphere with a certain number of handles, the number of handles being called the genus of the surface. In the case of a surface arising from an equation, it is natural to call this number the 'genus' of the equation. The genus for the Fermat equation with exponent n works out to be $(n-1)(n-2)/2$.

It turns out that the problem of finding rational solutions to the equation is closely related to the genus of the equation. The larger the genus, the more complicated is the geometry of the surface, and the harder it becomes to find rational points.

The simplest case is where the genus is 0, as is true for the 'Pythagorean' equations

$$x^2 + y^2 = k.$$

In this case, one of two results is possible. One possibility is that the equation has no rational points, like the equation

$$x^2 + y^2 = -1.$$

Alternatively, if there is a rational point, then it is possible to establish a one-to-one correspondence between all the rational numbers, t, and all the ra-

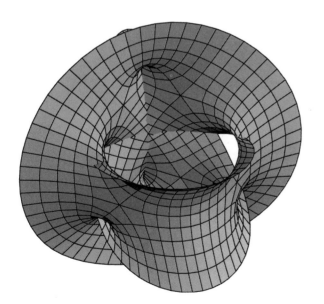
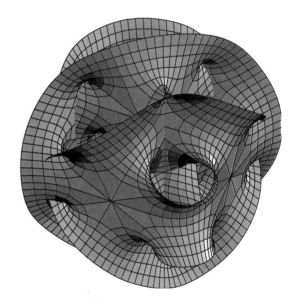

Surfaces generated by the Fermat equation $x^n + y^n = 1$, where x and y are regarded as complex variables. The figure on the left shows the surface for $n = 3$, the one on the right is for $n = 5$. Both figures were produced by the mathematics software system *Mathematica*, one of a number of sophisticated computer tools available to the modern mathematician.

tional points on the curve, as I did earlier for the circle (page 201). In this case, there are infinitely many rational solutions, and the *t*-correspondence gives a method for computing those solutions.

The case of curves of genus 1 is much more complicated. Curves determined by an equation of genus 1 are called elliptic curves, since they arise in the course of computing the length of part of an ellipse. Examples of elliptic curves are shown on this page. Elliptic curves have a number of nice properties that make them extremely useful in number theory. For example, some of the most-powerful known methods for factoring large integers into primes (on a computer) are based on the theory of elliptic curves.

As in the case of curves of genus 0, an elliptic curve may have no rational points. But if there is a rational point, then an interesting thing happens, as the English mathematician Lewis Mordell discovered in the early part of the century. Mordell showed that, although the number of rational points

may be either finite or infinite, there is always a *finite* set of rational points—they are called generators—such that all other rational points may be generated from them by a simple, explicit process. All that is required is some elementary algebra coupled with the drawing of lines that either are tangent to the curve or cut it at three points. Thus, even in the case where there are infinitely many rational points, there is a structure—a pattern—to handle them.

Of course, the genus 1 case is not in itself of particular interest if the goal is to prove Fermat's last theorem, which, if you discount the case of exponent $n = 3$, concerns equations of genus greater than 1. But, as a result of his investigations, in 1922 Mordell made an observation of tantalizing relevance to Fermat's last theorem: no one had ever found an equation of genus greater than 1 that has infinitely many rational solutions! In particular, all of the many equations that Diophantus had examined turned out to have genus 0 or 1. Mordell con-

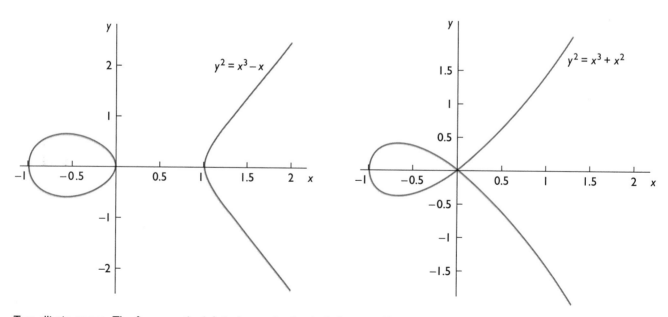

Two elliptic curves. The figure on the left is the graph of a single function. Even though it breaks apart into two separate pieces, mathematicians refer to it as a single 'curve'. The graph on the right crosses itself at the origin.

jectured that this was not just an accident, but that no equation of genus greater than 1 could have an infinite number of rational solutions.

In particular, Mordell's conjecture implied that, for each value of the exponent n greater than 2, the Fermat equation

$$x^n + y^n = 1$$

can have at most a finite number of rational solutions. Thus, a proof of the Mordell conjecture, while not proving Fermat's last theorem, would mark a significant development.

The Mordell conjecture was finally proved in 1983, by a young German mathematician named Gerd Faltings. Faltings had to combine a number of deep ideas in order to produce his proof. The first of these key ideas appeared in 1947 in the work of André Weil, who had been investigating whole-number solutions to equations with respect to finite arithmetics. Weil's basic question was, given a prime number p, how many whole-number solutions does an equation have modulo p? This question is obviously related to Fermat's last theorem, since, if there are no solutions modulo p, then there can be no solutions in the usual sense. By analogy with some results in topology, Weil formulated a number of technical conjectures concerning this problem. These conjectures were formulated in terms of what are known as 'algebraic varieties'—loosely speaking, sets of solutions not to a single equation but a whole system of equations. These conjectures were finally proved in 1975, by Pierre Deligne.

The second significant contribution to the proof of the Mordell conjecture arose from an analogy between ordinary equations, whose coefficients are numbers, and equations whose coefficients are rational functions, functions of the form $p(x)/q(x)$, where $p(x)$ and $q(x)$ are polynomials. This analogy is a very strong one, and many concepts and results in number theory have analogues for these 'function fields', as they are known. In particular, the Mordell conjecture has an analogue. When the Soviet mathematician Yuri Manin proved this analogue in 1963,

Gerd Faltings of Princeton University.

it provided additional evidence that the Mordell conjecture might turn out to be true.

A third ingredient of Falting's proof was the Shafarevich conjecture. Shortly before Manin obtained his result, his countryman Igor Shafarevich formulated a conjecture concerning the way in which information about the whole-number solutions to an equation can be 'pieced together' from solutions to certain other equations—namely, the equations that result when the original equation is interpreted in mod-p finite arithmetic for different primes p. In 1968, A. N. Parshin proved that the Shafarevich conjecture implies the Mordell conjecture.

Meanwhile, in 1966, a fourth contribution was made, when the American John Tate came up with yet another conjecture about algebraic varieties.

This proliferation of conjectures was a reflection of the growing understanding of the emerging structure. Generally, mathematicians only make a 'guess' public when they have some intuition to support it. In this case, all the different guesses were heading in the same direction. For his 1983 proof of the Mordell conjecture, Faltings first proved the Tate conjecture. Combining that with Deligne's results on the Weil conjectures, he was then able to establish the Shafarevich conjecture. Because of Parshin's 1968 result, this at once yielded the Mordell conjecture, and with it a proof of the fact that no Fermat equation can have an infinite number of solutions.

It was a marvellous illustration of the way increasing abstraction and a search for ever deeper patterns can lead to a proof of a quite concrete result—in this case, a result about whole-number solutions to simple equations.

Three years later, there was another major advance in our understanding of Fermat's last theorem. As with the proof of Mordell's conjecture, an intricate sequence of conjectures was involved, and once again elliptic curves played a significant role in the story.

In 1955, the Japanese mathematician Yutaka Taniyama had proposed that there should be a connection between elliptic curves and another well-understood—but not easily described—class of curves, known as modular curves. According to Taniyama, there should be a connection between any given elliptic curve and a modular curve, and this connection should 'control' many of the properties of the initial curve.

Taniyama's conjecture was made more precise in 1968 by André Weil, who showed how to determine the exact modular curve that should be connected to a given elliptic curve, and, in 1971, Goro Shimura demonstrated that Weil's procedure works for a very special class of equations. Taniyama's proposal became known as the Shimura–Taniyama (sometimes Shimura–Taniyama–Weil) conjecture.

So far, there was no known connection between this very abstract conjecture and Fermat's last theorem, and most mathematicians would have doubted that there was any connection at all. But, in 1986, Gerhard Frey, a mathematician from Saarbrucken, surprised everyone by finding a highly innovative link between the two.

Frey realized that, if there are whole numbers a, b, c, n such that $c^n = a^n + b^n$, then it seemed unlikely that one could understand the elliptic curve given by the equation

$$y^2 = x(x - a^n)(x + b^n)$$

in the way proposed by Taniyama. Following an appropriate reformulation of this observation by Jean-Pierre Serre, the American mathematician Kenneth Ribet proved conclusively that the existence of a counterexample to the Fermat's last theorem would in fact lead to the existence of an elliptic curve which could not be modular, and hence would contradict the Shimura–Taniyama conjecture. Thus, a proof of the Shimura–Taniyama conjecture would at once imply Fermat's last theorem.

This was tremendous progress. Now there was a definite structure to work with: the Shimura–Taniyama conjecture concerned geometric ob-

Kenneth Ribet of the University of California at Berkeley.

jects about which a great deal was known. Indeed, there was good reason to believe the result, and to suggest a way of setting about finding a proof. At least, the English mathematician Andrew Wiles saw a way to proceed.

Wiles had been fascinated with Fermat's last theorem since he was a child, when he had attempted to solve the problem using high school mathematics. Later, when he learned of Kummer's work as a student at Cambridge University, he tried again using the German's more sophisticated techniques. But when he realized how many mathematicians had tried and failed to solve the problem, he eventually gave up on it and concentrated on maistream contemporary number theory, in particular the theory of elliptic curves.

It was a fortuitous choice. For as soon as Ribet proved his astonishing and completely unexpected result, Wiles found himself an acknowledged world expert in the very techniques that could perhaps provide the elusive key to the proof of Fermat's last theorem. For the next seven years he concentrated all his efforts on trying to find a way to prove the Shimura–Taniyama Conjecture. By 1991, using powerful new methods developed by Barry Mazur, Matthias Flach, Victor Kolyvagin, and others, he

felt sure he could prove a special case of the Shimura–Taniyama Conjecture, which applied to elliptic curves of a particular kind. In 1993, after a further two years effort, he eventually succeeded in doing just that.

Believing that the class of elliptic curves for which his proof worked included those necessary to deduce Fermat's last theorem, in June 1993, at a small mathematics meeting in Cambridge, England, Wiles announced that he had succeeded. He had, he said, proved Fermat's last theorem.

He was wrong. By December of that year, he had to admit that his argument did not seem to work for the 'right' elliptic curves. Though everyone agreed that his achievement was one of the most significant advances in number theory in the twentieth century, it appeared that he was destined to follow in the footsteps of all those illustrious mathematicians of years gone by, including perhaps Fermat himself, who had dared rise to the challenge laid down in that tantalizing marginal note.

Several months of silence followed, while Wiles retreated to his Princeton home to try to make his argument work. In October 1994 he announced that, with the help of a former student, Richard Taylor of Cambridge University he had succeeded.

His proof—and this time everyone agreed it was correct—was given in two papers: a long one entitled "Modular Elliptic Curves and Fermat's Last Theorem," which contained the bulk of his argument, and a shorter, second paper co-authored with Taylor, entitled "Ring Theoretic Properties of Certain Hecke Algebras," which provided a key step Wiles used in his proof. The two papers together constitute the May 1995 issue of the prestigous research journal *Annals of Mathematics*.

Fermat's last theorem was a theorem at last.

The story of Fermat's last theorem is a marvellous illustration of humanity's never-ending search for knowledge and understanding. But it is much more than that. Mathematics is the only branch of science in which a precise, technical problem formulated in the seventeenth century, and having its origins in ancient Greece, remains as pertinent to-day as it did then. It is unique among the sciences in that a new development does not invalidate the previous theories, but builds on what has gone before. A long path leads from the Pythagorean theorem and Diophantus' *Arithmetic*, to Fermat's marginal comment, and on to the rich and powerful theory that finally settled the matter. A great many mathematicians have contributed to that development. They have lived (and are living) all over the world; they have spoken (and speak) many languages; most of them have never met. What has united them has been their love for mathematics. Over the years, each has helped the others, as new generations of mathematicians have adopted and adapted ideas of their predecessors. Separated by time, space, and culture, they have all contributed to a single enterprise. In this respect, perhaps mathematics can serve as an example to all humanity.

Postscript

There is more. A great deal more. The general themes outlined in the previous chapters cover just one small portion of present-day mathematics. For instance, there was no mention of Maxwell's electromagnetic theory, an entirely mathematical theory published in 1873. Our eyes can detect light, which is a form of electromagnetic radiation, and by turning on the radio or television we can verify that the air around us is teeming with radio waves, another form of electromagnetic radiation. The equations Maxwell discovered provide an extremely precise mathematical description of electromagnetic radiation, but beyond the mathematics the nature of an electromagnetic wave remains something of a mystery.

There was no mention of probability theory, developed by Girolamo Cardano, Blaise Pascal, Pierre de Fermat, and others in the seventeenth century, a theory that captures the patterns of random events. There was little or no mention of statistics, numerical analysis, computation theory, computational complexity theory, numerical analysis, approximation theory, dynamical systems theory, chaos theory, the theory of infinite numbers, game theory, the theory of voting systems, the theory of conflict, operations research, optimization theory, mathematical economics, the mathematics of finance, catastrophe theory, and weather forecasting. Little or no mention was made of applications of mathematics in physics, in engineering, in astronomy, in psychology, in biology, in chemistry, in ecology, and in aerospace science. Each one of these topics could have occupied a complete chapter in a book of this nature. So too could a number of other topics that I did not include in the above lists.

In writing this book, I wanted to convey some sense of the nature of mathematics, both the mathematics of today and its evolution over the course of history. But I did not want to serve up a vast smorgasbord of topics, each one being allotted a couple of pages. For all its many facets and its many connections to other disciplines and other walks of life, mathematics is very much a single whole. A mathematical study of any one phenomenon has many similarities to a mathematical study of any other. There is an initial simplification, when the key concepts are identified and isolated. Then those key concepts are analyzed in greater and greater depth, as the relevant patterns are discovered and investigated. There are attempts at axiomatization. The level of abstraction increases. Theorems are formulated and proved. Connections to other parts of mathematics are uncovered, or maybe just suspected. The theory is generalized, leading to the discovery of further similarities to—and connections with—other areas of mathematics. It is this overall structure of the field that I wanted to convey.

The particular topics I chose are all 'central' themes within mathematics. To some extent, they form the core of the broad subject we call mathematics. They are all included, to a greater or lesser extent, in any college or university undergraduate degree program in mathematics. In that sense, their selection was a natural one. But the fact is, I could have chosen any collection of five or six general areas and told the same story: that mathematics is the science of patterns, and those patterns can be found anywhere you care to look for them, in the physical universe, in the living world, or even in our own minds.

Further Readings

A comprehensive list of books that provide more detail of the topics discussed in the preceding pages would be long indeed. A sublist of those books that are even remotely accessible to the general reader would be extremely short. By and large, mathematicians have not written many books for the audience for which *Mathematics: The Science of Patterns* is designed. Apart from textbooks for the budding or professional mathematician, most mathematics books have been 'how to' books designed to develop mathematical 'skills'—arithmetic, elementary algebra, and the like. The following list is not exhaustive of the available books a typical reader of this volume might enjoy, but it does include most of the suitable ones known to the author.

In addition to the books listed, the reader of *Mathematics: The Science of Patterns* is unlikely to be disappointed with any book by Martin Gardner. Martin Gardner was for many years the author of the "Mathematical Games" section of *Scientific American*. Most of his regular columns have appeared in book form. The level of exposition is, on the whole, considerably less challenging than in *Mathematics: The Science of Patterns*. Some Gardner books are less mathematical than others, but all are well written. Don't worry too much about the title, just look for the author's name.

At about the same level as Gardner, that is to say, journalistic, are two books by Ivars Peterson, *The Mathematical Tourist* (Freeman, 1988) and *Islands of Truth* (Freeman, 1990). Peterson is a science journalist who writes for *Science News*. Good exposition at a fairly elementary level.

The Problems of Mathematics, by Ian Stewart (Oxford University Press, 1992). The spread of topics is larger than in *Mathematics: The Science of Patterns,* though there is less detail on each topic. Highly recommended, and clearly aimed at the same audience as *Mathematics: The Science of Patterns*.

Journey Through Genius: The Great Theorems of Mathematics, by William Dunham (Penguin Books, 1991). A personal selection of a small number of topics. A good exposition.

The Mathematical Experience, by Philip Davis and Reuben Hersh (Birkhäuser Boston, 1981). A marvellous account of what it is like to *do* mathematics as a professional.

To Infinity and Beyond: A Cultural History of the Infinite, by Eli Maor (Birkhäuser Boston, 1987). Though the overall theme is more restricted, readers of *Mathematics: The Science of Patterns* will find much that is familiar. A good exposition.

Geometry in Nature, by Vagn Lundsgaard Hansen (A K Peters, 1993). In many ways a possible companion to *Mathematics: The Science of Patterns,* this book focuses on the mathematical patterns found in physics.

The Art of Mathematics, by Jerry P. King (Plenum, 1992) is a beautifully written little book that sets out to describe what its title suggests. The author concentrates on the people who do mathematics rather than the mathematics they do.

Beyond the Third Dimension: Geometry, Computer Graphics, and Higher Dimensions, by Thomas Banchoff (Freeman, Scientific American Library, 1990). This excellent book takes the reader much further into the worlds of four and more dimensions than was possible in *Mathematics: The Science of Patterns*.

The Mathematical Universe, by William Dunham (John Wiley, 1994). A highly readable, alphabetic tour through mathematics.

Nature's Numbers, by Ian Stewart (Basic Books, 1995). An excellent book to start with. Marvellous, evocative prose with not a formula or equation in sight.

Five Golden Rules, by John L. Casti (John Wiley, 1996). About the same level as *The Science of Patterns,* but a different choice of topics.

The Parsimononious Universe, by Stefan Hildebrandt and Anthony Tromba (Springer-Verlag, 1996). This looks very reminiscent of a Scientific American Library Book, which is exactly how it began life, in 1984, under the title *Mathematics and Optimal Form.*

M. C. Escher: Visions of Symmetry, by Doris Schattschneider (Freeman, 1990). The author is a well known geometer, but this book concentrates on Escher's art, rather than the underlying mathematics.

The Visual Mind: Art and Mathematics, edited by Michele Emmer (The International Society for the Arts, Sciences and Technology, 1993). A collection of articles by leading authorities in mathematics and art. A fascinating book, having much in common with *Mathematics: The Science of Patterns.*

The Fourth Dimension and Non-Euclidean Geometry in Modern Art, by Linda Dalrymple Henderson (Princeton University Press, 1983). The author is an art historian, and the focus is on how artists have tried to come to terms with various advances in mathematics.

The Beauty of Fractals: Images of Complex Dynamical Systems, by H.-O. Peitgen and P. H. Richter (Springer-Verlag, 1986). A lavishly illustrated book, with brief explanations of the mathematics that lies behind the computer graphics displayed.

The Development of Logic, by William Kneale and Martha Kneale (Oxford University Press, 1962). This is a scholarly book that traces the development of mathematical logic from Greek times to the present era. Fascinating reading that provides a much more accurate picture than the brief description in Chapter 2 of *Mathematics: The Science of Patterns.*

Finally, two books for anyone who wants to get a taste of what is involved in doing modern mathematics:

Sets, Functions and Logic, by Keith Devlin (Chapman and Hall, 1992) is a textbook used by beginning college mathematics students. Readers of *Mathematics: The Science of Patterns,* should find it accessible.

Elementary Number Theory, by David Burton (Allyn and Bacon, 1980), is a college textbook on number theory that should be just about accessible to anyone who manages to get through *Sets, Functions and Logic.*

Sources of Illustrations

Line illustrations are by Ian Worpole and Publication Services.

Prologue
Opposite page 1: Wassily Kandinsky, *In the Black Square (Im schwarzem Viereck),* 1923, Solomon R. Guggenheim Museum, New York, N.Y. Gift of Solomon R. Guggenheim, 1937. *page 2:* Isaac Newton, *The Method of Fluxions and Infinite Series,* H. Woodfall, London, 1736. *page 4:* Office of Special Collections, New York Public Library. *page 5:* Johann Sebastian Bach, *Prelude in G Minor (early version),* BWV 535a Autograph, c. 1705–7. Stattsbibliothek zu Berlin-Preussischer Kulturbesitz, Muzikabteilung mit Mendlssohn-Archiv. *page 6:* From H. O. Peitgen and P. H. Richter, *The Beauty of Fractals,* Springer-Verlag, 1986. © H. Jürgens, H. O. Peitgen, and D. Saupe. *page 7:* Costa's surface. David Hoffman, James T. Hoffman, and Michael Callahan, Geometry Analysis Numeric and Graphics Center, University of Massachusetts, Amherst.

Chapter 1
page 8: Jasper Johns, *Numbers in Color,* 1958–9, Albright-Knox Art Gallery, Buffalo, N.Y. Gift of Seymour H. Knox, 1959. *page 11:* (top left and right) D. Schmandt-Besserat, University of Texas at Austin. Musée du Louvre, Département des Antiquités Orientales, Paris. (bottom left) D. Schmandt-Besserat, University of Texas at Austin. German Archaeology Institute, Division Baghdad. *page 13:* British Museum, Egyptian Antiquities. *page 18:* (top) Painting by Justus Ghent, Galleria Nazionale delle Marche, Palazzo Ducale, Urbino. Scala/Art Resource. (bottom) Euclid, *Elementia Geometricae,* Library of Congress, Washington, D.C., Rare Book Collection. *page 21:* Deutsches Museum, München.

page 23: Academie des Sciences, Inscriptions et Belles Lettres de Toulouse. *page 28:* Cray Research, Inc. *page 29:* Office of Special Collections, New York Public Library. *page 31:* Avakian/ Woodfin Camp and Assoc. *page 34:* Emanuel Handmann, *Portrait of Mathematician Leonhard Euler,* 1753, Öffentliche Kunstsammlung Basel, Kunstmuseum.

Chapter 2
page 36: Nicolaus Neufchatel, *Portrait of Johannes Neudorfer and his Son,* 1561, Bayerische Staatsgemäldesammlungen, München. Kunstdia-Archiv ARTOTHEK. *page 38:* Detail of Raphael, *School of Athens,* Vatican Palace. Erich Lessing/Art Resource. *page 43:* College Archives, University College, Cork, Ireland. *page 49:* Jean van Heijenoort, *From Frege to Gödel,* Harvard University Press, Cambridge, Mass., 1967. *page 49:* (top right) Motorola, Inc. *page 55:* Fernand Léger, *Still Life with Beer Mug,* 1921–2, Tate Gallery, London/Art Resource. *page 56:* Deutsches Museum, München. *page 58:* Keystone Press Agency. *page 59:* Jean van Heijenoort, *From Frege to Gödel,* Harvard University Press, Cambridge, Mass., 1967. *page 62:* Deutsches Museum, München. *page 63:* Archives, Institute of Advanced Study. Photo by Alfred Eisenstaedt, Life Magazine. © Time Warner. *page 64:* Stanford University News Service. *page 69:* Adapted from Frederick Mosteller et al., "Statistics: A guide to the unknown," in F. Mosteller and D. Wallace, *Deciding Authorship,* Wadsworth, 1989, page 122. *page 70:* Jerry Brendt.

Chapter 3
page 72: Marcel Duchamp, *Nude Descending a Staircase, No. 2,* 1912, Philadelphia Museum of Art. The Louise and Walter Arensberg Collection. *page 78:* Manfred Schroeder, *Fractals, Chaos and Power Laws,*

W. H. Freeman and Co., 1992. *page 83:* University Library, Cambridge, England, MS Add. 4004, Fol. 81 verso, The Syndics of Cambridge University Library. *page 84:* Anonymous, eighteenth century, Trinity College, Cambridge. Erich Lessing/Art Resource. *page 85:* Pierre-Louis Moreau de Maupertuis, "Les Lois du mouvement et du repos, déduites d'un principe de métaphysique," *Mémoires de l'Académie de Berlin,* 1746. *page 87:* Emilio Segre Visual Archives, Center for History of Physics, American Institute of Physics. *page 88:* Deutsches Museum, München. *page 93:* NASA.

Chapter 4

page 104: Rice Pereira, *Oblique Progression,* 1948, Whitney Museum of Art. *page 108:* Chip Clark. *page 115:* Öffentliche Bibliothek der Univerität, Basel. *page 117:* (top left) Domenico Fetti, *Archimedes,* Gemäldegalerie Alte Meister, Dresden. (bottom right): Frans Hals, Louvre/Art Resource. *page 120:* "The Lawnmower Man," courtesy Allied Vision Lane Pringle Productions. © 1992. All rights reserved. Computer animation by Angel Studios, Carlsbad, Calif. Directed by Brett Leonard. *page 125:* Boyer-Viollet. *page 129:* M. C. Escher, *Circle Limit III,* 1959, Cordon Art, Baarn, Holland. *page 130:* (top) Adapted from W. Berlinghoff and K. Grant, *A Mathematical Sampler,* Ardsley House Publishers, 1992. (bottom) Albrecht Dürer, *Institutiones Geometricae,* 1525. *page 131:* The Master of the Barberini Panels (Umbrian-Florentine), *The Annunciation,* 1450, National Gallery of Art, Washington, D.C. Samuel H. Kress Collection. *page 132:* M. C. Escher, *Man with Cuboid,* 1958, Cordon Art, Baarn, Holland. *page 136:* (left) Albrecht Dürer, *St. Jerome in his Study,* 1514. (right) Martin Kemp, *The Science of Art,* Yale University Press, New Haven, Conn., 1990, page 61. *page 140:* (top) Gareth White. (bottom) Attilio Pierelli. *page 141:* Max Weber, *Interior*

of the Fourth Dimension, 1913, National Gallery of Art, Washington, D.C. *page 142:* Thomas Banchoff and Charles Strauss.

Chapter 5

page 144: Hans van Lemmen. *page 146:* Nuridsany et Perenou/Science Source/Photo Researchers. *page 150:* P. Dupuy, *La vie d'Evariste Galois,* Paris, 1903. *page 153:* Hugh Sitton/Tony Stone Images. *page 158:* William Neill. *page 161:* Adapted from *What's Happening in the Mathematical Sciences, Vol. 1,* American Mathematical Society, 1993, page 25. *page 162:* Chip Clark. *page 163:* Victoria and Albert Museum. Bridgeman Art Library/Art Resource. *page 164 (bottom right)* and *page 166:* Ivars Peterson, *Islands of Truth,* W. H. Freeman and Co., 1990. *page 168:* (bottom) Paul Steinhardt, University of Pennsylvania. *page 169:* A. R. Kortan, AT&T Bell Laboratories. *page 170:* M. C. Escher, *Symmetry Drawing E2,* Cordon Art, Baarn, Holland. *page 171:* Stan Sherer.

Chapter 6

page 172: M. C. Escher, *Three Spheres 1,* 1945, Cordon Art, Baarn, Holland. *page 174:* London Transport Museum. *page 179:* M. C. Escher, *Möbius Strip II,* 1963, The Escher Foundation, Haags Gemeentenmuseum, The Hague. *page 182:* James T. Hoffman, Geometry Analysis Numeric and Graphics Center, University of Massachusetts, Amherst. *page 186:* Chip Clark. *page 188:* Roger-Viollet. *page 191:* Stephanie Rausser. *page 194:* Thomas Banchoff and Nicholas Thompson. *page 198:* Adapted from Keith Devlin, *The New Golden Age,* Penguin Books, 1987, page 242. *page 199:* Nicholas R. Cozzarelli, University of California at Berkeley, and Steven A. Wasserman, University of Texas at Austin. *page 203:* Wolfram Research. *page 205:* Donald Albers. *page 206:* Catherine Karnow. *page 207:* Science Source/Photo Researchers.

Index